Technology and the Politics of Knowledge

THE INDIANA SERIES
IN THE PHILOSOPHY OF TECHNOLOGY

Don Ihde, *general editor*

TECHNOLOGY

AND THE

POLITICS OF KNOWLEDGE

Edited by
Andrew Feenberg and Alastair Hannay

Indiana University Press

Bloomington and Indianapolis

The paper used in this publication meets the minimum requirements of American
National Standard for Information Sciences—Permanence of Paper for Printed
Library Materials, ANSI Z39.48-1984.

Manufactured in the United States of America

Library of Congress Cataloging-in-Publication Data

Technology and the politics of knowledge / edited by Andrew
Feenberg and Alastair Hannay.
p.   cm. — (Indiana series in the philosophy of technology)
Includes bibliographical references and index.
ISBN 0-253-32154-9. — ISBN 0-253-20940-4 (pbk.)
1. Technology—Philosophy.   I. Feenberg, Andrew.   II. Hannay,
Alastair.   III. Series.
T14.T387   1995
601—dc20   94-27789

2 3 4 5 00 99 98 97

Contents

Preface

THE PHILOSOPHY OF technology is at a turning point. Not long ago the very label lacked seriousness. Technology was for technologists and if philosophers had anything to say to practitioners, their professional concern was with goals not means. Indeed the prospect of a philosophy of technology had to await the dissolution of the widely held belief that philosophy was interested in technology only to condemn it.

In recent years that has all changed. Environmentalism has undermined the old confidence in the objectivity of scientific-technical expertise. In a world of continuing war, prejudice, illness, hunger, and poverty we can no longer afford to be complacent about technology, and a superior disregard of this ever-expanding sphere of human life has now given way to broad-based concern. The ready availability of computer technology has no doubt played its part in changing attitudes. Considerable progress in historical, sociological, and cultural studies of technology has made available a large body of literature on every aspect of the subject and philosophers have not been slow to appropriate it.

Where formerly the discussion of technology was discouraged among philosophers by the cult of value-neutrality and the programmatic subservience of philosophy to science in matters of knowledge, there are now several philosophical traditions which stress the role of values in the growth of knowledge. Thus a clearing has opened within which fundamental issues of political and social philosophy can be raised in their relevance to the rapid technological change going on around us. Indeed, concern with technology has become unavoidable in the context of discussions of Habermas, Foucault, Heidegger, and feminism, where the question of its nature and consequences frequently looms large.

A not uncommon response to the facts that prompt concern with technology is despair, or else a utopian optimism. Traces of neither will be found in these pages, nor is there any common consensus. But the reader will note with what surprising regularity the authors propose that philosophical reflection and democratic discussion should play a larger role in shaping the technological environment. The reader will also notice how pervasive the theme of technology is, how diverse the areas of everyday life in which it plays a part, and, not least, how long the problems to which it gives rise have been waiting to be posed.

This collection presents a generous sampling of the new philosophy of technology. Bringing together an unexpectedly wide array of authors and subject matter, it makes evident the maturity that the field has attained. The seven sections into which the volume is divided review the classic contributions of the Frankfurt School and Heidegger as well as innovative methods in the social construction of science and technology, such recent issues as the feminist critique of reproductive technologies, and the impact of technology on multiculturalism, the body, and postmodern politics.

Acknowledgments

FOR PERMISSION TO include previously published materials the editors are grateful to the authors, editors, and publishers in question. The following acknowledgments are due: to Scandinavian University Press for permission to reprint Andrew Feenberg, "Subversive Rationalization: Technology, Power, and Democracy"; Langdon Winner, "Citizen Virtues in a Technological Order"; Albert Borgmann, "The Moral Significance of the Material Culture"; Don Ihde, "Image Technologies and Traditional Culture"; Yaron Ezrahi, "Technology and the Civil Epistemology of Democracy"; Helen E. Longino, "Knowledge, Bodies, and Values: Reproductive Technologies and Their Scientific Context"; Pieter Tijmes, "The Archimedean Point and Eccentricity: Hannah Arendt's Philosophy of Science and Technology"; and Paul Dumouchel, "Gilbert Simondon's Plea for a Philosophy of Technology," all of which originally appeared in *Inquiry* 35, nos. 3/4 (1992), © 1992 Scandinavian University Press; *Research in Philosophy and Technology* and JAI Press for permission to reprint Steven Vogel, "New Science, New Nature: The Habermas-Marcuse Debate Revisited," which originally appeared in vol. 11 (1991), © 1991 by JAI Press, Inc.; to Kluwer Academic Publishers for permission to reprint Robert B. Pippin, "On the Notion of Technology as Ideology," which originally appeared in *Sociology of the Sciences Yearbook* (1993), pp. 91–113, © 1993 by Kluwer Academic Publishers, and Tom Rockmore, "Heidegger on Technology and Democracy," which originally appeared in Langdon Winner (ed.), *Democracy in a Technological Society*, pp. 187–205, © 1992 Kluwer Academic Publishers; to *Feminist Studies* for permission to reprint Donna Haraway, "Situated Knowledges: The Science Question in Feminism and the Privilege of Partial Perspective," which originally appeared in vol. 14, no. 3 (1988), © 1988 by Feminist Studies, Inc.; and to Editions Dupuis for permission to reproduce an extract from album no. 13 of "Gaston," *Lagaffe mérite des baffes*, by André Franquin, © 1979 by Franquin and Editions Dupuis; it appeared in an earlier version of Bruno Latour, "A Door Must Be Either Open or Shut: A Little Philosophy of Techniques," which appeared in Bruno Latour, *Le Cléf de Berlin*, Paris: Editions la Découverte, 1993.

PART I

Technology as Ideology

IS TECHNOLOGY SOCIALLY and ethically neutral, a product of rational problem solving, or is it, as the Frankfurt School claimed, a kind of materialized ideology, a prop of the established society? The three chapters of this section discuss different aspects of this question.

In "Subversive Rationalization: Technology, Power, and Democracy" Andrew Feenberg updates the Frankfurt School approach in terms of the new constructivist sociology of technology. He argues that technology is not governed by economic and technical rationality, nor is it determining for society. On the contrary it is technology that is determined in its meaning and normative content by the social world in which it is embedded. But then technology, like other institutions, ought to be subject to conscious social control. An anti-foundationalist philosophy of technology can help to restore the eroded initiative of the democratic process in determining the future of technologically advanced societies.

Steven Vogel's "New Science, New Nature: The Habermas-Marcuse Debate Revisited" reviews the controversy over the relativity of scientific-technical knowledge in the Frankfurt School. At his most utopian, Marcuse called for a new non-alienated science and technology in a pacified society. Habermas rejects any such reconciliation with nature as romantic and instead links science and technology to a generic human interest in control. Vogel teases out contradictions in Habermas's position, but Marcuse too confronts unresolvable paradoxes because he posits an independent nature as the object of a socially determined science and technology. To recognize the social construction of nature without reserve is finally to accept the human responsibility for science and the natural world.

Robert B. Pippin's "On the Notion of Technology as Ideology" evaluates a wide range of arguments for the technology-as-ideology thesis, from Heidegger's theory of "enframing," through Horkheimer and Adorno's critique of enlightenment, to Habermas's theory of the technological "colonization of the lifeworld." He concludes that all of these theories share the questionable assumption that there is a "true interest" in a better design or application of technology lurking beneath the surface of modern social life. Pippin claims, on the contrary, that normative confusion, acquisitiveness, instrumentalism, and ceaseless self-expansion are defining for modernity and are merely reflected in technology as we know it. The problem is not with technology, with the control of machines or their place in social life, but with modernity as such.

1

Subversive Rationalization

Technology, Power, and Democracy[1]

Andrew Feenberg

The Limits of Democratic Theory

TECHNOLOGY IS ONE of the major sources of public power in modern societies. So far as decisions affecting our daily lives are concerned, political democracy is largely overshadowed by the enormous power wielded by the masters of technical systems: corporate and military leaders, and professional associations of groups such as physicians and engineers. They have far more to do with control over patterns of urban growth, the design of dwellings and transportation systems, the selection of innovations, our experience as employees, patients, and consumers, than all the governmental institutions of our society put together.

Marx saw this situation coming in the middle of the nineteenth century. He argued that traditional democratic theory erred in treating the economy as an extra-political domain ruled by natural laws such as the law of supply and demand. He claimed that we will remain disenfranchised and alienated so long as we have no say in industrial decision making. Democracy must be extended from the political domain into the world of work. This is the underlying demand behind the idea of socialism.

Modern societies have been challenged by this demand for over a century. Democratic political theory offers no persuasive reason of principle to reject it. Indeed, many democratic theorists endorse it.[2] What is more, in a number of countries, socialist parliamentary victories or revolutions have brought parties to power dedicated to achieving it. Yet today we do not appear to be much closer to democratizing industrialism than in Marx's time.

This state of affairs is usually explained in one of the following two ways. On the one hand, the commonsense view argues that modern technology is incompatible with workplace democracy. Democratic theory cannot reasonably press for reforms that would destroy the economic foundations of society. For evidence, consider the Soviet case: although they were socialists, the communists did not democratize industry, and the current democratization of Soviet society extends only to the factory gate. At least in the ex-Soviet Union, everyone can agree on the need for authoritarian industrial management.

On the other hand, a minority of radical theorists claim that technology is not responsible for the concentration of industrial power. That is a political matter, due to the victory of capitalist and communist elites in struggles with the underlying population. No doubt modern technology lends itself to authoritarian administration, but in a different social context it could just as well be operated democratically.

In what follows, I will argue for a qualified version of this second position, somewhat different from both the usual Marxist and social-democratic formulations. The qualification concerns the role of technology, which I see as *neither* determining nor as neutral. I will argue that modern forms of hegemony are based on the technical mediation of a variety of social activities, whether it be production or medicine, education or the military, and that, consequently, the democratization of our society requires radical technical as well as political change.

This is a controversial position. The commonsense view of technology limits democracy to the state. By contrast, I believe that unless democracy can be extended beyond its traditional bounds into the technically mediated domains of social life, its use-value will continue to decline, participation will wither, and the institutions we identify with a free society will gradually disappear.

Let me turn now to the background to my argument. I will begin by presenting an overview of various theories that claim that insofar as modern societies depend on technology, they require authoritarian hierarchy. These theories presuppose a form of technological determinism which is refuted by historical and sociological arguments I will briefly summarize. I will then present a sketch of a non-deterministic theory of modern society I call "critical theory of technology." This alternative approach emphasizes contextual aspects of technology ignored by the dominant view. I will argue that technology is not just the rational control of nature; both its development and impact are intrinsically social. I will then show that this view undermines the customary reliance on efficiency as a criterion of technological development. That conclusion, in turn, opens broad possibilities of change foreclosed by the usual understanding of technology.

Dystopian Modernity

Max Weber's famous theory of rationalization is the original argument against industrial democracy. The title of this paper implies a provocative reversal of Weber's conclusions. He defined rationalization as the increasing role of calculation and control in social life, a trend leading to what he called the "iron cage" of bureaucracy. "Subversive" rationalization is thus a contradiction in terms.[3]

Once the traditionalist struggle against rationalization has been defeated, further resistance in a Weberian universe can only reaffirm irrational life-forces against routine and drab predictability. This is not a democratic program but a romantic anti-dystopian one, the sort of thing that is already foreshadowed in Dostoievsky's *Notes from Underground* and various back-to-nature ideologies.

My title is meant to reject the dichotomy between rational hierarchy and irrational protest implicit in Weber's position. If authoritarian social hierarchy is truly a contingent dimension of technical progress, as I believe, and not a technical necessity, then there must be an alternative way of rationalizing society that democratizes rather than centralizes control. We need not go underground or native to preserve threatened values such as freedom and individuality.

But the most powerful critiques of modern technological society follow directly in Weber's footsteps in rejecting this possibility. I am thinking of Heidegger's formulation of "the question of technology" and Ellul's theory of "the technical phenomenon."[4] According to these theories, we have become little more than objects of technique, incorporated into the mechanism we have created. As Marshall McLuhan once put it, technology has reduced us to the "sex organs of machines." The only hope is a vaguely evoked spiritual renewal that is too abstract to inform a new technical practice.

These are interesting theories, but I have time to do little more than pay tribute to their contribution to opening a space of reflection on modern technology. Instead, to advance my own argument, I will concentrate on their principal flaw, the identification of technology in general with the specific technologies that have developed in the last century in the West. These are technologies of conquest that pretend to an unprecedented autonomy; their social sources and impacts are hidden. I will argue that this type of technology is a particular feature of our society and not a universal dimension of "modernity" as such.

Technological Determinism

Determinism rests on the assumption that technologies have an autonomous functional logic that can be explained without reference to society. Technology is presumably social only through the purpose it serves, and purposes are in the mind of the beholder. Technology would thus resemble science and mathematics by its intrinsic independence of the social world.

Yet unlike science and mathematics, technology has immediate and powerful social impacts. It would seem that society's fate is at least partially dependent on a nonsocial factor which influences it without suffering a reciprocal influence. This is what is meant by "technological determinism."

The dystopian visions of modernity I have been describing are deterministic. If we want to affirm the democratic potentialities of modern industrialism, we will therefore have to challenge their deterministic premises. These I will call the thesis of unilinear progress and the thesis of determination by the base. Here is a brief summary of these two positions.

1. Technical progress appears to follow a unilinear course, a fixed track, from less to more advanced configurations. Although this conclusion seems obvious from a backward glance at the development of any familiar technical object, in fact it is based on two claims of unequal plausibility: first, that technical progress proceeds from lower to higher levels of development; and second, that that development follows a single sequence of necessary stages. As we will see, the first claim is independent of the second and not necessarily deterministic.

2. Technological determinism also affirms that social institutions must adapt to the "imperatives" of the technological base. This view, which no doubt has its source in a certain reading of Marx, is now part of the common sense of the social sciences.[5] Below, I will discuss one of its implications in detail: the supposed "trade-off" between prosperity and environmental ideology.

These two theses of technological determinism present decontextualized, self-generating technology as the unique foundation of modern society. Determinism thus implies that our technology and its corresponding institutional structures are universal, indeed planetary, in scope. There may be many forms of tribal society, many feudalisms, even many forms of early capitalism, but there is only one modernity and it is exemplified in our society, for good or ill. Developing societies should take note: as Marx once said, calling the attention of his backward German compatriots to British advances: "*De te fabula narratur*"—of you the tale is told.[6]

Constructivism

The implications of determinism appear so obvious that it is surprising to discover that neither of its two theses can withstand close scrutiny. Yet contemporary sociology of technology undermines the first thesis of unilinear progress while historical precedents are unkind to the second thesis of determination by the base.

Recent constructivist sociology of technology grows out of new social studies of science. These studies challenge our tendency to exempt scientific theories from the sort of sociological examination to which we submit non-scientific beliefs. They affirm the "principle of symmetry," according to which all contending beliefs are subject to the same type of social explanation regardless of their truth or falsity.[7] A similar approach to technology rejects the usual assumption that technologies succeed on purely functional grounds.

Constructivism argues that theories and technologies are underdetermined by scientific and technical criteria. Concretely, this means two things: first, there is generally a surplus of workable solutions to any given problem, and social actors make the final choice among a batch of technically viable options; and second, the problem-definition often changes in the course of solution. The latter point is the more conclusive, but also the more difficult, of the two.

Two sociologists of technology, Pinch and Bijker, illustrate it with the early history of the bicycle.[8] The object we take to be a self-evident "black box" actually started out as two very different devices, a sportsman's racer and a utilitarian transportation vehicle. The high front wheel of the sportsman's bike was necessary at the time to attain high speeds, but it also caused instability. Equal-sized wheels made for a safer but less exciting ride. These two designs met different needs and were in fact different technologies with many shared elements. Pinch and Bijker call this original ambiguity of the object designated as a "bicycle," "interpretive flexibility."

Eventually the "safety" design won out, and it benefited from all the later advances that occurred in the field. In retrospect, it seems as though the high-wheelers were a clumsy and less efficient stage in a progressive development leading through the old "safety" bicycle to current designs. In fact the high-wheeler and the safety bicycle shared the field for years, and neither was a stage in the other's development. The high-wheeler represents a possible alternative path of bicycle development that addressed different problems at the origin.

Determinism is a species of Whig history which makes it seem as though the end of the story was inevitable from the very beginning by projecting the abstract technical logic of the finished object back into the past as a cause of development. That approach confuses our understanding of the past and stifles the imagination of a different future. Constructivism can open up that future, although its practitioners have hesitated so far to engage the larger social issues implied in their method.[9]

Indeterminism

If the thesis of unilinear progress falls, the collapse of the notion of determination by the technological base cannot be far behind. Yet it is still frequently invoked in contemporary political debates.

I shall return to these debates later in this chapter. For now, let us consider the remarkable anticipation of current attitudes in the struggle over the length of the workday and child labor in mid-nineteenth-century England. Factory owners and economists denounced regulation as inflationary; industrial production supposedly required children and the long workday. One member of parliament declared that regulation is "a false principle of humanity, which in the end is certain to defeat itself." He went on to argue that the new rules were so radical as to constitute "in principle an argument to get rid of the whole system of factory labor."[10] Similar protestations are heard today on behalf of industries threatened with what they call environmental "Luddism."

Yet what actually happened once the regulators succeeded in imposing limitations on the workday and expelling children from the factory? Did the violated imperatives of technology come back to haunt them? Not at all. Regulation led to an intensification of factory labor that was incompatible with the earlier conditions in any case. Children ceased to be workers and were redefined socially as learners and consumers. Consequently, they entered the labor market with higher levels of skill and discipline that were soon presupposed by technological design. As a result no one is nostalgic for a return to the good old days when inflation was held down by child labor. That is simply not an option.

This example shows the tremendous flexibility of the technical system. It is not rigidly constraining but, on the contrary, can adapt to a variety of social demands. This conclusion should not be surprising given the responsiveness of technology to social redefinition discussed previously. It means that technology is just another dependent social variable, albeit an increasingly important one, and not the key to the riddle of history.

Determinism, I have argued, is characterized by the principles of unilinear progress and determination by the base; if determinism is wrong, then technology research must be guided by the following two contrary principles. In the first place, technological development is not unilinear but branches in many directions and could reach generally higher levels along more than one different track. And, second, technological development is not determining for society but is overdetermined by both technical and social factors.

The political significance of this position should also be clear by now. In a society where determinism stands guard on the frontiers of democracy, indeterminism cannot but be political. If technology has many unexplored potentialities, no technological imperatives dictate the current social hierarchy. Rather, technology is a scene of social struggle, a "parliament of things," on which civilizational alternatives contend.

Interpreting Technology

In the remainder of this chapter I would like to present several major themes of a non-determinist approach to technology. The picture sketched so far implies a significant change in our definition of technology. It can no longer be considered as a collection of devices, or, more generally, as the sum of rational means. These are tendentious definitions that make technology seem more functional and less social than in fact it is.

As a social object, technology ought to be subject to interpretation like any other cultural artifact, but it is generally excluded from humanistic study. We are assured that its essence lies in a technically explainable function rather than a hermeneutically interpretable meaning. At most, humanistic methods might illuminate extrinsic aspects of technology, such as packaging and advertising, or popular reactions to controversial innovations such as nuclear power or surrogate motherhood. Technological determinism draws its force from this attitude. If one ignores most of the connections between technology and society, it is no wonder that technology then appears to be self-generating.

Technical objects have two hermeneutic dimensions that I call their *social meaning* and their *cultural horizon.*[11] The role of social meaning is clear in the case of the bicycle introduced above. We have seen that the construction of the bicycle was controlled in the first instance by a contest of interpretations: was it to be a sportsman's toy or a means of transportation? Design features such as wheel size also served to signify it as one or another type of object.[12]

It might be objected that this is merely an initial disagreement over goals with no hermeneutic significance. Once the object is stabilized, the engineer has the last word on its nature, and the humanist interpreter is out of luck. This is the view of most engineers and managers; they readily grasp the concept of "goal" but they have no place for "meaning."

In fact the dichotomy of goal and meaning is a product of functionalist professional culture, which is itself rooted in the structure of the modern economy. The concept of "goal" strips technology bare of social contexts, focusing engineers and managers on just what they need to know to do their job.

A fuller picture is conveyed, however, by studying the social role of the technical object and the lifestyles it makes possible. That picture places the abstract notion of "goal" in its concrete social context. It makes technology's contextual causes and consequences visible rather than obscuring them behind an impoverished functionalism.

The functionalist point of view yields a decontextualized temporal cross-section in the life of the object. As we have seen, determinism claims implausibly to be able to get from one such momentary configuration of the object to the next on purely technical terms. But in the real world all sorts of unpredictable attitudes crystallize around technical objects and influence later design changes. The engineer may think these are extrinsic to the device he or she is working on, but they are its very substance as a historically evolving phenomenon.

These facts are recognized to a certain extent in the technical fields themselves, especially in computers. Here we have a contemporary version of the dilemma of the bicycle discussed above. Progress of a generalized sort in speed, power, and memory goes on apace while corporate planners struggle with the question of what it is all for. Technical development does not point definitively toward any particular path. Instead, it opens branches, and the final determination of the "right" branch is not within the competence of engineering, because it is simply not inscribed in the nature of the technology.

I have studied a particularly clear example of the complexity of the relation between the technical function and meaning of the computer in the case of French videotex.[13] Called "Teletel," this system was designed to bring France into the Information Age by giving telephone subscribers access to databases. Fearing that consumers would reject anything resembling office equipment, the telephone company attempted to redefine the computer's social image; it was no longer to appear as a calculating device for professionals but was to become an informational network for all.

The telephone company designed a new type of terminal, the Minitel, to look and feel like an adjunct to the domestic telephone. The telephonic disguise suggested to some users that they ought to be able to talk to each other on the network. Soon the Minitel underwent a further redefinition at the hands of these users, many of whom employed it primarily for anonymous on-line chatting with other users in the search for amusement, companionship, and sex.

Thus the design of the Minitel invited communications applications which the company's engineers had not intended when they set about improving the flow of information in French society. Those applications, in turn, connoted the Minitel as a means of personal encounter, the very opposite of the rationalistic project for which it was originally created. The "cold" computer became a "hot" new medium.

At issue in the transformation is not only the computer's narrowly conceived technical function, but the very nature of the advanced society it makes possible. Does networking open the doors to the Information Age where, as rational consumers hungry for data, we pursue strategies of optimization? Or is it a postmodern technology that emerges from the breakdown of institutional and sentimental stability, reflecting, in Lyotard's words, the "atomization of society into flexible networks of language games"?[14] In this case technology is not merely the servant of some pre-defined social purpose; it is an environment within which a way of life is elaborated.

In sum, differences in the way social groups interpret and use technical objects are not merely extrinsic but make a difference in the nature of the objects themselves. *What* the object *is* for the groups that ultimately decide its fate determines what it *becomes* as it is redesigned and improved over time. If this is true, then we can understand technological development only by studying the sociopolitical situation of the various groups involved in it.

Technological Hegemony

In addition to the sort of assumptions about individual technical objects we have been discussing so far, that situation also includes broader assumptions about social values. This is where the study of the cultural horizon of technology comes in. This second hermeneutic dimension of technology is the basis of modern forms of social hegemony; it is particularly relevant to our original question concerning the inevitability of hierarchy in technological society.

As I will use the term, hegemony is a form of domination so deeply rooted in social life that it seems natural to those it dominates. One might also define it as that aspect of the distribution of social power which has the force of culture behind it.

The term *horizon* refers to culturally general assumptions that form the unquestioned background to every aspect of life.[15] Some of these support the prevailing hegemony. For example, in feudal societies, the "chain of being" established hierarchy in the fabric of God's universe and protected the caste relations of the society from challenge. Under this horizon, peasants revolted in the name of the king, the only imaginable source of power. Rationalization is our modern horizon, and technological design is the key to its effectiveness as the basis of modern hegemonies.

Technological development is constrained by cultural norms originating in economics, ideology, religion, and tradition. We discussed earlier how assumptions about the age-composition of the labor force entered into the design of nineteenth-century production technology. Such assumptions seem so natural and obvious that they often lie below the threshold of conscious awareness.

This is the point of Herbert Marcuse's important critique of Weber.[16] Marcuse shows that the concept of rationalization confounds the control of labor by management with control of nature by technology. The search for control of nature is generic, but management arises only against a specific social background, the capitalist wage system. Workers have no immediate interest in output in this system, unlike earlier forms of farm and craft labor, since their wage is not essentially linked to the income of the firm. Control of human beings becomes all-important in this context.

Through mechanization, some of the control functions are eventually transferred from human overseers and parcelized work practices to machines. Machine design is thus socially relative in a way that Weber never recognized, and the "technological rationality" it embodies is not universal but particular to capitalism. In fact, it is the horizon of all the existing industrial societies, communist as well as capitalist, insofar as they are managed from above. (In a later section, I discuss a generalized application of this approach in terms of what I call the "technical code.")

If Marcuse is right, it ought to be possible to trace the impress of class relations in the very design of production technology, as has indeed been shown by such Marxist students of the labor process as Harry Braverman and David Noble.[17] The assembly line offers a particularly clear instance because it achieves traditional management goals, such as de-skilling and pacing work, through technical design. Its technologically enforced labor discipline increases productivity and profits by increasing control. However, the assembly line appears as technical progress only in a specific social context. It would not be perceived as an advance in an economy based on workers' cooperatives in which labor discipline was more self-imposed than imposed from above. In such a society, a different technological rationality would dictate different ways of increasing productivity.[18]

This example shows that technological rationality is not merely a belief, an ideology, but is effectively incorporated into the structure of machines. Machine design mirrors back the social factors operative in the prevailing rationality. The fact that the argument for the social relativity of modern technology originated in a Marxist context has obscured its most radical implications. We are not dealing here with a mere critique of the property system, but have extended the force of that critique down into the technical "base." This approach goes well beyond the old economic distinction between capitalism and socialism, market and plan. Instead, one arrives at a very different distinction between societies in which power rests on the technical mediation of social activities and those that democratize technical control and, correspondingly, technological design.

Double Aspect Theory

The argument to this point might be summarized as a claim that social meaning and functional rationality are inextricably intertwined dimensions of technology. They are not ontologically distinct, for example, with meaning in the observer's mind and rationality in the technology proper. Rather they are "double aspects" of the same underlying technical object, each aspect revealed by a specific contextualization.

Functional rationality, like scientific-technical rationality in general, isolates objects from their original context in order to incorporate them into theoretical or functional systems. The institutions that support this procedure—such as laboratories and research centers—themselves form a special context with their own practices and links to various social agencies and powers. The notion of "pure" rationality arises when the work of decontextualization is not itself grasped as a social activity reflecting social interests.

Technologies are selected by these interests from among many possible configurations. Guiding the selection process are social codes established by the cultural and political struggles that define the horizon under which the technology will fall. Once introduced, technology offers a material validation of the cultural horizon to which it has been preformed. I call this the "bias" of technology: apparently neutral, functional rationality is enlisted in support of a hegemony. The more technology society employs, the more significant is this support.

As Foucault argues in his theory of "power/knowledge," modern forms of oppression are not so much based on false ideologies as on effective techniques "encoded" by the dominant hegemony to reproduce the system.[19] So long as that act of choice remains hidden, the deterministic image of a technically justified social order is projected.

The legitimating effectiveness of technology depends on unconsciousness of the cultural-political horizon under which it was designed. A recontextualizing critique of technology can uncover that horizon, demystify the illusion of technical necessity, and expose the relativity of the prevailing technical choices.

The Social Relativity of Efficiency

These issues appear with particular force in the environmental movement today. Many environmentalists argue for technical changes that would protect nature and in the process improve human life as well. Such changes would enhance efficiency in broad terms by reducing harmful and costly side effects of technology. However, this program is very difficult to impose in a capitalist society. There is a tendency to deflect criticism from technological processes to products and people, from a priori prevention to a posteriori cleanup. These preferred strategies are generally costly and reduce efficiency under the horizon of the given technology. This situation has political consequences.

Restoring the environment after it has been damaged is a form of collective consumption, financed by taxes or higher prices. These approaches dominate public awareness. This is why environmentalism is generally perceived as a cost involving trade-offs, and not as a rationalization increasing overall efficiency. But in a modern society, obsessed by economic well-being, that perception is damning. Economists and businessmen are fond of explaining the price we must pay in inflation and unemployment for worshiping at Nature's shrine instead of Mammon's. Poverty awaits those who will not adjust their social and political expectations to technology.

This trade-off model has environmentalists grasping at straws for a strategy. Some hold out the pious hope that people will turn from economic to spiritual values in the face of the mounting problems of industrial society. Others expect enlightened dictators to bite the bullet of technological reform even if a greedy populace shirks its duty. It is difficult to decide which of these solutions is more improbable, but both are incompatible with basic democratic values.[20]

The trade-off model confronts us with dilemmas—environmentally sound technology vs. prosperity, workers' satisfaction and control vs. productivity, etc.—where what we need are syntheses. Unless the problems of modern industrialism can be solved in ways that both enhance public welfare and win public support, there is little reason to hope that they will ever be solved. But how can technological reform be reconciled with prosperity when it places a variety of new limits on the economy?

The child-labor case shows how apparent dilemmas arise on the boundaries of cultural change, specifically where the social definition of major technologies is in transition. In such situations, social groups excluded from the original design network articulate their unrepresented interests politically. New values which the outsiders believe would enhance their welfare appear as mere ideology to insiders who are adequately represented by the existing designs.

This is a difference of perspective, not of nature. Yet the illusion of essential conflict is renewed whenever major social changes affect technology. At first, satisfying the demands of new groups after the fact has visible costs and, if it is done clumsily, will indeed reduce efficiency until better designs are found. But, usually, better designs can be found and what appeared to be an insuperable barrier to growth dissolves in the face of technological change.

This situation indicates the essential difference between economic exchange and technique. Exchange is all about trade-offs: more of A means less of B. But the aim of

technical advance is precisely to avoid such dilemmas by elegant designs that opti-
mize several variables at once. A single cleverly conceived mechanism may correspond to many different social demands, one structure to many functions.[21] Design is
not a zero-sum economic game but an ambivalent cultural process that serves a multiplicity of values and social groups without necessarily sacrificing efficiency.

The Technical Code

That these conflicts over social control of technology are not new can be seen
from the interesting case of the "bursting boilers."[22] Steamboat boilers were the first
technology the U.S. government subjected to safety regulation. Over five thousand
people were killed or injured in hundreds of steamboat explosions from 1816, when
regulation was first proposed, to 1852, when it was actually implemented. Is this many
casualties or few? Consumers evidently were not too alarmed to continue traveling by
riverboat in ever-increasing numbers. Understandably, the shipowners interpreted this
as a vote of confidence and protested the excessive cost of safer designs. Yet politicians
also won votes demanding safety.

The accident rate fell dramatically once technical improvements were mandated.
Legislation would hardly have been necessary to achieve this outcome had it been
technically determined. But in fact boiler design was relative to a social judgment
about safety. That judgment could have been made on strictly market grounds, as the
shippers wished, or politically, with differing technical results. In either case, those
results *constitute* a proper boiler. What a boiler "is" was thus defined through a long
process of political struggle culminating finally in uniform codes issued by the American Society of Mechanical Engineers.

This example shows just how technology adapts to social change. What I call the
"technical code" of the object mediates the process. That code responds to the cultural
horizon of the society at the level of technical design. Quite down-to-earth technical
parameters, such as the choice and processing of materials, are *socially* specified by
the code. The illusion of technical necessity arises from the fact that the code is thus
literally "cast in iron" or "set in concrete" as the case may be.[23]

Conservative anti-regulatory social philosophies are based on this illusion. They
forget that the design process always already incorporates standards of safety and environmental compatibility; similarly, all technologies support some basic level of user
or worker initiative. A properly made technical object simply *must* meet these standards to be recognized as such. We do not treat conformity as an expensive add-on, but
regard it as an intrinsic production cost. Raising the standards means altering the definition of the object, not paying a price for an alternative good or ideological value as
the trade-off model holds.

But what of the much discussed cost-benefit ratio of design changes such as those
mandated by environmental or other similar legislation? These calculations have some

application to transitional situations, before technological advances responding to new values fundamentally alter the terms of the problem. But all too often the results depend on economists' very rough estimates of the monetary value of such things as a day of trout fishing or an asthma attack. If made without prejudice, these estimates may well help to prioritize policy alternatives. But one cannot legitimately generalize from such policy applications to a universal theory of the costs of regulation.

Such fetishism of efficiency ignores our ordinary understanding of the concept which alone is relevant to social decision making. In that everyday sense, efficiency concerns the narrow range of values that economic actors routinely affect by their decisions. Unproblematic aspects of technology are not included. In theory one can decompose any technical object and account for each of its elements in terms of the goals it meets, whether it be safety, speed, reliability, etc., but in practice no one is interested in opening the "black box" to see what is inside.

For example, once the boiler code is established, such things as the thickness of a wall or the design of a safety valve appear as essential to the object. The cost of these features is not broken out as the specific "price" of safety and compared unfavorably with a more efficient but less secure version of the technology. Violating the code in order to lower costs is a crime, not a trade-off. And since all further progress takes place on the basis of the new safety standard, soon no one looks back to the good old days of cheaper, insecure designs.

Design standards are controversial only while they are in flux. Resolved conflicts over technology are quickly forgotten. Their outcomes, a welter of taken-for-granted technical and legal standards, are embodied in a stable code and form the background against which economic actors manipulate the unstable portions of the environment in the pursuit of efficiency. The code is not varied in real-world economic calculations but treated as a fixed input.

Anticipating the stabilization of a new code, one can often ignore contemporary arguments that will soon be silenced by the emergence of a new horizon of efficiency calculations. This is what happened with boiler design and child labor; presumably, the current debates on environmentalism will have a similar history, and we will some-day mock those who object to cleaner air as a "false principle of humanity" that violates technological imperatives.

Non-economic values intersect the economy in the technical code. The examples we are dealing with illustrate this point clearly. The legal standards that regulate workers' economic activity have a significant impact on every aspect of their lives. In the child-labor case, regulation helped to widen educational opportunities with consequences that are not primarily economic in character. In the riverboat case, Americans gradually chose high levels of security, and boiler design came to reflect that choice. Ultimately, this was no trade-off of one good for another, but a non-economic decision about the value of human life and the responsibilities of government.

Technology is thus not merely a means to an end; technical design standards define major portions of the social environment, such as urban and built spaces, work-places, medical activities and expectations, life patterns, and so on. The economic significance of technical change often pales beside its wider human implications in framing a way of life. In such cases, regulation defines the cultural framework *of* the economy; it is not an act *in* the economy.

The Consequences of Technology

The theory sketched here suggests the possibility of a general reform of technol-ogy. But dystopian critics object that the mere fact of pursuing efficiency or technical effectiveness already does inadmissible violence to human beings and nature. Univer-sal functionalization destroys the integrity of all that is. As Heidegger argues, an "ob-jectless" world of mere resources replaces a world of "things" treated with respect for their own sake as the gathering places of our manifold engagements with "being."[24]

This critique gains force from the actual perils with which modern technology threatens the world today. But my suspicions are aroused by Heidegger's famous con-trast between a dam on the Rhine and a Greek chalice. It would be difficult to find a more tendentious comparison. No doubt modern technology is immensely more de-structive than any other. And Heidegger is right to argue that means are not truly neu-tral, that their substantive content affects society independent of the goals they serve. But this content is not *essentially* destructive; rather, it is a matter of design and social insertion.

In another text, Heidegger shows us a jug "gathering" the contexts in which it was created and functions. There is no reason why modern technology cannot also "gather" its multiple contexts, albeit with less romantic pathos. This is in fact one way of interpreting contemporary demands for such things as environmentally sound tech-nology, applications of medical technology that respect human freedom and dignity, urban designs that create humane living spaces, production methods that protect workers' health and offer scope for their intelligence, and so on. What are these de-mands if not a call to reconstruct modern technology, so that it gathers a wider range of contexts to itself rather than reducing its natural, human, and social environment to mere resources?

Heidegger would not take these alternatives very seriously, because he reifies modern technology as something separate from society, as an inherently contextless force aiming at pure power. If this is the "essence" of technology, reform would be merely extrinsic. But at this point Heidegger's position converges with the very Prometheanism he rejects. Both depend on the narrow definition of technology that, at least since Bacon and Descartes, has emphasized its destiny to control the world to the exclusion of its equally essential contextual embeddedness. I believe that this defi-nition reflects the capitalist environment in which modern technology first developed.

The exemplary modern master of technology is the entrepreneur, single-mindedly focused on production and profit. The enterprise is a radically decontextualized platform for action, without the traditional responsibilities for persons and places that went with technical power in the past. It is the autonomy of the enterprise that makes it possible to distinguish so sharply between intended and unintended consequences, between goals and contextual effects, and to ignore the latter.

The narrow focus of modern technology meets the needs of a particular hegemony; it is not a metaphysical condition. Under that hegemony technological design is unusually decontextualized and destructive. It is that hegemony that is called to account, not technology *per se*, when we point out that today technical means form an increasingly threatening life-environment. It is that hegemony, as it has embodied itself in technology, that must be challenged in the struggle for technological reform.

The "Essence" of Technology

Heidegger rejects any merely social diagnosis of the ills of technological societies and claims that the source of their problems dates back at least to Plato, that modern societies merely realize a *telos* immanent in Western metaphysics from the beginning. His originality consists in pointing out that the ambition to control being is itself a way of being and hence subordinate at some deeper level to an ontological dispensation beyond human control. Heidegger's demand for a new response to the challenge of this dispensation is so shrouded in obscurity that no one has ever been able to give it a concrete content. The overall effect of his critique is to condemn human agency, at least in modern times, and to confuse essential differences between types of technological development.

This confusion has a historical aspect. Heidegger is perfectly aware that technical activity was not "metaphysical" in his sense until recently. He must therefore sharply distinguish modern technology from all earlier forms of technique, obscuring the many real connections and continuities. I would argue, on the contrary, that what is new about modern technology can be understood only against the background of the traditional technical world from which it developed. Furthermore, the saving potential of modern technology can be realized only by recapturing certain traditional features of technique. Perhaps this is why theories that treat modern technology as a unique phenomenon lead to such pessimistic conclusions.

Modern technology differs from earlier technical practices through significant shifts in emphasis rather than generically. There is nothing unprecedented in its chief features, such as the reduction of objects to raw materials, the use of precise measurement and plans, the technical control of some human beings by others, large scales of operation. It is the centrality of these features that is new, and of course the consequences of that are truly without precedent.

What does a broader historical picture of technology show? The privileged dimensions of modern technology appear in a larger context that includes many currently subordinated features that were defining for it in former times. For example, until the generalization of Taylorism, technical life was essentially about the choice of a vocation. Technology was associated with a way of life, with specific forms of personal development, virtues, etc. Only the success of capitalist de-skilling finally reduced these human dimensions of technique to marginal phenomena.

Similarly, modern management has replaced the traditional collegiality of the guilds with new forms of technical control. Just as vocational investment in work continues in certain exceptional settings, so collegiality survives in a few professional or cooperative workplaces. Numerous historical studies show that these older forms are incompatible not so much with the "essence" of technology as with capitalist economics. Given a different social context and a different path of technical development, it might be possible to recover these traditional technical values and organizational forms in new ways in a future evolution of modern technological society.

Technology is an elaborate complex of related activities that crystallizes around tool-making and -using in every society. Matters such as the transmission of techniques or the management of its natural consequences are not extrinsic to technology *per se* but are dimensions of it. When, in modern societies, it becomes advantageous to minimize these aspects of technology, that too is a way of accommodating it to a certain social demand, not the revelation of its pre-existing "essence." Insofar as it makes sense to talk about an essence of technology at all, it must embrace the whole field revealed by historical study, and not only a few traits ethnocentrically privileged by our society.

Conclusion: Subversive Rationalization

For generations faith in progress was supported by two widely held beliefs: that technical necessity dictates the path of development, and that the pursuit of efficiency provides a basis for identifying that path. I have argued here that both these beliefs are false, and that, furthermore, they are ideologies employed to justify restrictions on opportunities to participate in the institutions of industrial society. I conclude that we can achieve a new type of technological society which can support a broader range of values. Democracy is one of the chief values a redesigned industrialism could better serve.

What does it mean to democratize technology? The problem is not primarily one of legal rights but of initiative and participation. Legal forms may eventually routinize claims that are asserted at first informally, but the forms will remain hollow unless they emerge from the experience and needs of individuals resisting a specifically technological hegemony.

That resistance takes many forms, from union struggles over health and safety in nuclear power plants to community struggles over toxic waste disposal to political demands for regulation of reproductive technologies. These movements alert us to the need to take technological externalities into account and demand design changes responsive to the enlarged context revealed in that accounting.

Such technological controversies have become an inescapable feature of contemporary political life, laying out the parameters for official "technology assessment."[25] They prefigure the creation of a new public sphere embracing the technical background of social life, and a new style of rationalization that internalizes unaccounted costs borne by "nature," i.e., something or -body exploitable in the pursuit of profit. Here respect for nature is not antagonistic to technology but enhances efficiency in broad terms.

As these controversies become commonplace, surprising new forms of resistance and new types of demands emerge alongside them. Networking has given rise to one among many such innovative public reactions to technology. Individuals who are incorporated into new types of technical networks have learned to resist through the net itself in order to influence the powers that control it. This is not a contest for wealth or administrative power, but a struggle to subvert the technical practices, procedures, and designs structuring everyday life.

Again, the example of the Minitel can serve as a model of this new approach. In France, the computer was politicized as soon as the government attempted to introduce a highly rationalistic information system to the general public. Users "hacked" the network in which they were inserted and altered its functioning, introducing human communication on a vast scale where only the centralized distribution of information had been planned.

It is instructive to compare this case to the movements of AIDS patients.[26] Just as a rationalistic conception of the computer tends to occlude its communicative potentialities, so in medicine, caring functions have become mere side effects of treatment, which is itself understood in exclusively technical terms. Patients become objects of this technique, more or less "compliant" to management by physicians. The incorporation of thousands of incurably ill AIDS patients into this system destabilized it and exposed it to new challenges.

The key issue was access to experimental treatment. In effect, clinical research is one way in which a highly technologized medical system can care for those it cannot yet cure. But, until quite recently, access to medical experiments has been severely restricted by paternalistic concern for patients' welfare. AIDS patients were able to open up access because the networks of contagion in which they were caught were paralleled by social networks that were already mobilized around gay rights at the time the disease was first diagnosed.

Instead of participating in medicine individually as objects of a technical practice, they challenged it collectively and politically. They "hacked" the medical system and turned it to new purposes. Their struggle represents a counter-tendency to the technocratic organization of medicine, an attempt at a recovery of its symbolic dimension and caring functions.

As in the case of the Minitel, it is not obvious how to evaluate this challenge in terms of the customary concept of politics. Nor do these subtle struggles against the growth of silence in technological societies appear significant from the standpoint of the reactionary ideologies that contend noisily with capitalist modernism today. Yet the demand for communication these movements represent is so fundamental that it can serve as a touchstone for the adequacy of our concept of politics to the technological age.

These resistances, like the environmental movement, challenge the horizon of rationality under which technology is currently designed. Rationalization in our society responds to a particular definition of technology as a means to the goal of profit and power. A broader understanding of technology suggests a very different notion of rationalization, based on responsibility for the human and natural contexts of technical action. I call this "subversive rationalization" because it requires technological advances that can be made only in opposition to the dominant hegemony. It represents an alternative to both the ongoing celebration of technocracy triumphant and the gloomy Heideggerian counterclaim that 'Only a God can save us' from techno-cultural disaster.[27]

But is subversive rationalization in this sense socialist? There is certainly room for discussion of the connection between this new technological agenda and the old idea of socialism. I believe there is significant continuity. In socialist theory, workers' lives and dignity stood for the larger contexts which modern technology ignores. The destruction of their minds and bodies in the workplace was viewed as a contingent consequence of capitalist technical design. The implication that socialist societies might design a very different technology under a different cultural horizon was perhaps given only lip service, but at least it was formulated as a goal.

We can make a similar argument today over a wider range of contexts in a broader variety of institutional settings with considerably more urgency. I am inclined to call such a position socialist and to hope that in time it can replace the image of socialism projected by the failed communist experiment.

More important than this terminological question is the substantive point I have been trying to make. Why has democracy not been extended to technically mediated domains of social life despite a century of struggles? Is it because technology excludes democracy, or because it has been used to block it? The weight of the argument supports the second conclusion. Technology can support more than one type of technological civilization, and may someday be incorporated into a more democratic society than ours.

Notes and References

1. This paper expands a presentation of my book *Critical Theory of Technology* (New York: Oxford University Press, 1991), delivered at the American Philosophical Association, 28 Dec. 1991.

2. See, e.g., Joshua Cohen and Joel Rogers, *On Democracy: Toward a Transformation of American Society* (Harmondsworth: Penguin, 1983); Frank Cunningham, *Democratic Theory and Socialism* (Cambridge: Cambridge University Press, 1987).

3. Max Weber, *The Protestant Ethic and the Spirit of Capitalism*, trans. T. Parsons (New York: Scribners, 1958), pp. 181–82.

4. Martin Heidegger, *The Question Concerning Technology*, trans. W. Lovitt (New York: Harper & Row, 1977); Jacques Ellul, *The Technological Society*, trans. J. Wilkinson (New York: Vintage, 1964).

5. Richard W. Miller, *Analyzing Marx: Morality, Power and History* (Princeton: Princeton University Press, 1984), pp. 188–95.

6. Karl Marx, *Capital* (New York: Modern Library, 1906), p. 13.

7. See, e.g., David Bloor, *Knowledge and Social Imagery* (Chicago: University of Chicago Press, 1991), pp. 175–79. For a general presentation of constructivism, see Bruno Latour, *Science in Action* (Cambridge, MA: Harvard University Press, 1987).

8. Trevor Pinch and Wiebe Bijker, "The Social Construction of Facts and Artefacts: Or How the Sociology of Science and the Sociology of Technology Might Benefit Each Other," *Social Studies of Science*, no. 14, 1984.

9. See Langdon Winner's blistering critique of the characteristic limitations of the position, entitled "Upon Opening the Black Box and Finding It Empty: Social Constructivism and the Philosophy of Technology," *The Technology of Discovery and the Discovery of Technology: Proceedings of the Sixth International Conference of the Society for Philosophy and Technology* (Blacksburg, VA: The Society for Philosophy and Technology, 1991).

10. *Hansard's Debates, Third Series: Parliamentary Debates 1830–1891*, vol. LXXIII, 1844 (22 Feb.–22 April), pp. 1123 and 1120.

11. A useful starting-point for the development of a hermeneutics of technology is offered by Paul Ricoeur in "The Model of the Text: Meaningful Action Considered as a Text," P. Rabinow and W. Sullivan (eds.), *Interpretive Social Science: A Reader* (Berkeley: University of California Press, 1979).

12. Michel de Certeau used the phrase "rhetorics of technology" to refer to the representations and practices that contextualize technologies and assign them a social meaning. De Certeau chose the term *rhetoric* because that meaning is not simply present at hand but communicates a content that can be articulated by studying the connotations technology evokes. See the special issue of *Traverse*, no. 26, Oct. 1982, entitled *Les Rhétoriques de la Technologie*, and, in that issue, especially Marc Guillaume's article, *Téléspectres* (pp. 22–23).

13. Andrew Feenberg, "From Information to Communication: The French Experience with Videotex," in Martin Lea (ed.), *The Social Contexts of Computer Mediated Communication* (London: Harvester Wheatsheaf, 1992).

14. Jean-François Lyotard, *La Condition Postmoderne* (Paris: Editions de Minuit, 1979), p. 34.

15. For an approach to social theory based on this notion (called, however, *doxa*, by the author), see Pierre Bourdieu, *Outline of a Theory of Practice*, trans. R. Nice (Cambridge: Cambridge University Press, 1977), pp. 164–70.

16. Herbert Marcuse, "Industrialization and Capitalism in the Work of Max Weber," in *Negations*, trans. J. Shapiro (Boston: Beacon Press, 1968).

17. Harry Braverman, *Labor and Monopoly Capital* (New York: Monthly Review, 1974); David Noble, *Forces of Production* (New York: Oxford University Press, 1984).

18. Bernard Gendron and Nancy Holstrom, "Marx, Machinery and Alienation," *Research in Philosophy and Technology* 2 (1979).

19. Foucault's most persuasive presentation of this view is *Discipline and Punish*, trans. A. Sheridan (New York: Vintage Books, 1979).

20. See, e.g., Robert Heilbroner, *An Inquiry into the Human Prospect* (New York: Norton, 1975). For a review of these issues in some of their earliest formulations, see Andrew Feenberg, "Beyond the Politics of Survival," *Theory and Society* 7 (1979), no. 3.

21. This aspect of technology, called concretization, is explained in Gilbert Simondon, *La Mode d'Existence des Objets Techniques* (Paris: Aubier, 1958), chapter 1.

22. John G. Burke, "Bursting Boilers and the Federal Power," in M. Kranzberg and W. Davenport (eds.), *Technology and Culture* (New York: New American Library, 1972).

23. The technical code expresses the "standpoint" of the dominant social groups at the level of design and engineering. It is thus relative to a social position without for that matter being a mere ideology or psychological disposition. As I will argue in the last section of this chapter, struggle for socio-technical change can emerge from the subordinated standpoints of those dominated within technological systems. For more on the concept of standpoint epistemology, see Sandra Harding, *Whose Science? Whose Knowledge?* (Ithaca: Cornell University Press, 1991).

24. The texts by Heidegger discussed here are, in order, *The Question Concerning Technology*, op. cit.; and "The Thing," *Poetry, Language, Thought*, trans. A. Hofstadter (New York: Harper & Row, 1971).

25. Alberto Cambrosio and Camille Limoges, "Controversies as Governing Processes in Technology Assessment," in *Technology Analysis and Strategic Management* 3 (1991), no. 4.

26. For more on the problem of AIDS in this context, see Andrew Feenberg, "On Being a Human Subject: Interest and Obligation in the Experimental Treatment of Incurable Disease," *The Philosophical Forum* 23 (1992), no. 3.

27. "Only a God Can Save Us Now," Martin Heidegger interviewed in *Der Spiegel*, trans. D. Schendler, *Graduate Philosophy Journal* 6 (1977), no. 1.

2

New Science, New Nature

The Habermas-Marcuse Debate Revisited

Steven Vogel

THE TRADITION OF German Western Marxism has had a problem about nature from the beginning. For orthodox Marxism there was no difficulty—Marxist social theory represented the application of the successful methods of natural science to the realms of history and society. Once this view was rejected, however, and with it Engels's dream of an overarching "dialectics of nature" into which Marx's account of social theory fit as simply a part, it was no longer clear what viewpoint Marxist theory should take about nature or the science that investigated it. Two contradictory impulses have been warring among Western Marxists ever since, both with roots in Lukács. One asserts that "nature is a social category,"[1] and hence that our views of nature, including our natural science, are ideological, forming part of an intellectual superstructure subject to change "after the revolution." The other asserts that Marxism is only a social theory, not a theory of nature, and that if we want to know what nature is like contemporary natural science offers a perfectly adequate (and non-ideological) answer. Scientific claims about nature are in themselves socially neutral, for this second view; it is not natural science itself but its misapplication to the social realm that is ideological. "When the ideal of scientific knowledge is applied to nature," writes Lukács, "it simply furthers the progress of science. But when it is applied to society it turns out to be an ideological weapon of the bourgeoisie."[2]

The first view has its roots in Western Marxism's Hegelian epistemological position, which insists on the constitutive role of the (social) subject in the object of its knowledge, and so rejects the objectivism that underlies contemporary natural science as a symptom of reification; the second has its roots in Western Marxism's materialism, which wants to ground social theory in concrete reality, and which fears the apparently "idealist" and relativist implications of the "nature is a social category" claim. The two positions are not reconcilable, I believe. The result has been a series of antinomies about the status of nature and the ideological character of natural science that have plagued Western Marxism from its inception.

My purpose in this chapter is to offer some evidence for this thesis, by using one important episode from the history of Western Marxism as an illustration—the dispute between Herbert Marcuse and Jürgen Habermas, played out in a series of articles and books of the 1960s and early 1970s, about the "ideological" character of science and technology.[3] This dispute marked a kind of watershed in Western Marxist discussions of science, in which Marcuse's defense of the classical Frankfurt School position criticizing science as the "domination of nature" began to give way in favor of Habermas's sophisticated reformulation of Lukács's claim that the problem lay in the misapplication of natural science to social questions, not in natural science itself.[4] I argue that *neither* position is tenable, and that each suffers from a set of fundamental problems whose origin lies in the ambivalence about the social character of scientific knowledge mentioned above. In the first and second sections I present Marcuse's account of science as ideology and Habermas's critique of it, going on to argue in the third and fourth sections that Habermas's critique is vitiated by a fundamental antinomy within his own positive theory of science. Although this would seem to leave Marcuse's position standing, the fifth section shows that the same considerations that work against Habermas's view in fact reveal Marcuse's theoretical weaknesses as well: each figure is unable to assimilate conclusions about the active character of our relations with nature to his prior commitment to a "materialism" that sees nature as independent of the human, and the result in each case is conceptual failure.

Marcuse and the "New Science"

Although the earlier Frankfurt School texts, especially Horkheimer and Adorno's *Dialectic of Enlightenment*, had taken a deeply pessimistic view of the modern project of science and technology, connecting it with an attempt to "dominate nature" that inevitably ended in "enlightenment's" own self-destruction,[5] Marcuse's position of the 1950s was considerably more ambiguous. Technology appeared to him to provide, if not a guarantee of human freedom, then at least a necessary condition for it. In *Eros and Civilization* Marcuse argued that the development of automation had finally made it possible to reduce the labor time necessary for the satisfaction of human needs to a bare minimum, and in so doing had opened up the possibility of an end to the repression labor had made necessary. His vision of a new society thus depended essentially on the virtual abolition of labor (and scarcity) by technology.[6] Freedom is identified with the liberation of the repressed "inner nature" that Horkheimer and Adorno had described, but this is now seen as being made possible precisely by the completion of technological development, which Marcuse provocatively describes not as the abolition but as the consummation of labor's alienation—"The more complete the alienation of labor, the greater the potential of freedom: total automation would be the optimum. It is the sphere outside labor which defines freedom and fulfillment."[7] Hence whereas labor on this view is inevitably a realm of unfreedom, automation allows us to minimize it and so opens up a sphere of play, of polymorphous perversity, *external* to labor—the only sphere in which freedom is possible.

This is a paradoxical conclusion, because it makes the realm of freedom dependent on the continuation of a realm of necessity in itself explicitly associated with toil, repression, and domination. Assumed here is a sharp dichotomy between the technological sphere of labor on the one hand and the erotic sphere of play on the other that seems to reproduce, not to transcend, the bourgeois view of work as something to be avoided as much as possible. Marcuse sees this problem, which accounts for the ambiguity in the treatment of technology in *Eros and Civilization.*[8] He is consistently driven toward a claim that the framework of that book cannot easily handle—that the liberation of human powers which a nonrepressive society would make possible must necessarily transform the very character of labor, and with it technology.[9] Both realms, that is to say, that of work as well as that of play, need liberation; or, rather, it is the very distinction between the two that must be abolished.[10]

This means first of all that the easy optimism about contemporary technology's ability to overcome scarcity and produce a realm of freedom has to be rejected; to the extent that the technological project is in itself inimical to liberation it cannot serve as a foundation for it. "The machine is *not neutral*," writes Marcuse; "technology is always a historical-social *project*: in it is projected what a society and its ruling interests intend to do with men and things."[11] It means secondly as well that our concept of a liberated society must be expanded to include a "liberated technology": a technology that could found human liberation would have to be a "new" technology, one that viewed its own project and the nature it engaged with in a new way. These two positions became, in the 1960s, the center of Marcuse's discussions of science and technology.[12]

Thus in *One-Dimensional Man* one finds a critique of "technological rationality" that sees contemporary science and technology as by their very nature yoked to capitalism and domination. Technology's very "neutrality," its apparent indifference to the political or social purposes to which it is put, serves at a deeper level, Marcuse argues, to tie it to a society in which substantive critical discourse about values is systematically prevented.[13] By treating the external world as merely the matter of instrumental manipulation, and as subject only to formal mathematical laws, science and technology deny the objective reality of ethical, aesthetic, or political values, leaving only the value of control as unquestionable.

To break the link between science and domination would thus require an entirely new conception of scientific method, and with it a new technology; it is this possibility that Marcuse raises in certain celebrated, and highly utopian, passages in *One-Dimensional Man.*

Science, *by virtue of its own method* and concepts, has projected and promoted a universe in which the domination of nature has remained linked to the domination of man—a link which tends to be fatal to this universe as a whole. . . . If this is the case, then the change in the direction of progress, which might sever this fatal link, would also affect the very structure of science—the scientific project. Its hypotheses, without losing their rational character, would develop in an entirely different experimental context (that of a pacified world); consequently, science would arrive at essentially different concepts of nature and establish essentially different facts.[14]

Here a position only hinted at in *Eros and Civilization* is explicitly asserted: that science and technology are not simply the material foundation of the erotic "liberation of nature" that Marcuse's utopian thought calls for, but rather must themselves be *transformed* by that liberation, to the extent of radically changing their very methods and even results. If the domination of humans and that of nature are essentially linked, then a nondominative society would require a nondominative view of nature, and with it a new sort of science and technology that would instantiate this view. *One-Dimensional Man* does not say much about what a New Science would look like; the book is still marked by a pessimism about the possibility of alternative social forces arising that could counter the loss of the "second" or critical dimension Marcuse sees as the central fact of contemporary mass society. As the 1960s went on, Marcuse's pessimism began to lift, as he started to discern in third-world revolutionary movements, in the emerging counterculture, and then above all in the burgeoning of feminism the concrete development of a new sensibility that might indeed serve as the foundation for a new approach to nature.[15]

Thus in later works such as *An Essay on Liberation* and *Counterrevolution and Revolt* he allowed himself to speculate a bit about what this new sensibility might involve. No longer treating nature as a mere object to be controlled and manipulated, the new sensibility would now treat it, he says, "as a *subject* in its own right—a subject with which to live in a common human universe."[16] Not only humans but nature thus deserve liberation—"nature, too," Marcuse writes, "awaits the revolution!"[17] "In sharp contrast to the capitalist exploitation of nature," he asserts, the new approach "would be nonviolent, nondestructive: oriented on the life-enhancing, sensuous, aesthetic qualities inherent in nature."[18] The result would be "the reconstruction of reality," a transformation of the world by "a science and technology released from their service to destruction and exploitation" that would explicitly see their task as part of a "collective *practice of creating an environment* . . . in which the nonaggressive, erotic, receptive faculties of man, in harmony with the consciousness of freedom, strive for the pacification of man and nature."[19]

The notion of "reconstructing reality" is an important one: Marcuse emphasizes that his position is not a Luddite rejection of technology so much as a call for a new technology that does not leave nature as it is, but rather transforms it in the interest of liberation—ours and nature's own.[20] "No longer condemned to compulsive aggressiveness and repression in the struggle for existence," he writes, "individuals would be able to create a technical and natural environment which would no longer perpetuate violence, ugliness, ignorance, and brutality."[21] Such a transformation would distinguish itself from what in *One-Dimensional Man* Marcuse called the "repressive mastery" of nature characteristic of contemporary technology, turning rather to a "liberating mastery"[22] that paradoxically frees nature in its very transformation of it, allowing nature (in a phrase of Adorno's that Marcuse quotes) "on the poor earth to become what perhaps it would like to be."[23] This idea had already arisen in *Eros and Civilization*—that an "aesthetic" or "erotic" attitude toward nature transforms the things of nature in such a way as to allow them to realize their own inherent telos—"The things of nature become free to be what they are. But to be what they are they *depend* on the erotic attitude: they receive their *telos* only in it."[24]

Thus by its connection with a new technology, the "new science" Marcuse posits would be associated not only with a new method and with new results, but literally with a new world. "Released from the bondage to exploitation," he writes in *An Essay on Liberation*, "the imagination, sustained by the achievements of science, could turn its productive power to the radical reconstruction of experience and the universe of experience"; "the rational transformation of the world could then lead to a reality formed by the aesthetic sensibility of man."[25] Taken seriously, of course, such claims are strong indeed. With the notion of a nonrepressive society bringing into being a "new science" whose very epistemological foundations seem radically distinct from those of the natural science we know, Marcuse clearly wants to relativize science and technology to the social order, and in a way that has a fundamental ontological import. To say that a nondominative science would literally "establish different facts," and that with its associated technology it would "reconstruct reality," is to return to Lukács's thesis that "nature is a social category" with a vengeance: nature itself seems to change when society does.[26] It is this conclusion above all, with its apparent fall into an out-and-out idealism, that Habermas wants to challenge. Let us turn now to his critique of Marcuse's utopian speculations.

Habermas: Science as a Universal Species-Interest

Habermas's critique derives directly from the theory of "knowledge-constitutive interests" that formed the central element of the epistemological position he elaborated in the course of the 1960s.[27] With this theory Habermas reintroduced a Lukácsian dualism into Western Marxism by asserting the existence of two fundamental and mutually irreducible modes of human action—"work" and "interaction"—that he argued form the "quasi-transcendental" basis for the possibility of knowledge in the areas of natural science and social theory respectively.[28] These modes of action are built into the structure of the species as such—as Homo sapiens, we are necessarily characterized from the beginning by the dual projects of finding ways to provide ourselves with the physical necessities of life on the one hand and of interacting with our fellow creatures in the communicative practices that make up the social order on the other. In work, we use tools to change the world around us to satisfy our needs; in interaction, we use language to interpret our actions, our fellow-beings, and ourselves.[29]

The two modes of action thus reflect fundamental "human interests": built into the species, Habermas argues, are interests both in the prediction and control of nature (the "technical" interest) and in the achievement of mutual understanding on the basis of undistorted communication (the "communicative" interest). Our knowledge of the world is not, for Habermas, the result of some contemplative apprehension of an objective reality, but rather is in its very essence "interested"—which is to say, the objects we claim to know are objects "constituted" by us, in accordance with the fundamental interests, through our actions.[30]

Thus Habermas's dualism has an epistemological meaning. The natural sciences, on his view, do not investigate "reality as such," but rather that segment of the world constituted by the human interest in prediction and control of the environment; the *Geisteswissenschaften*, marked not by empirical methods but by hermeneutic ones, investigate a different segment, constituted by the human interest in achieving mutual understanding. In this way positivism's claim that knowledge of society must show the same methodological structure as the knowledge achieved by natural science, as well as its even stronger claim that the latter sort of knowledge attains an "objectivity" that *Verstehen* or other methods do not, both appear as untenable.[31]

With this elegant move, Habermas is able to reject the positivist identification of natural scientific knowledge with knowledge as such without at the same time leaving the undeniably successful natural sciences without any epistemological justification at all. The result, however, is to make Marcuse's call for a liberation of nature based on a "New Science" impossible. For science and technology, "interested" though they may be, still turn out on Habermas's view not to be connected to any particular social project, but rather to a project of the species *as a whole*: to a trans-social "species-interest" in prediction and control of nature.[32] Science is not, on this view, the ideological reflection of a social order based on domination, but rather simply the most recent form of a practice and a knowledge that have been characteristic of all human societies from the start—and that have their roots in the very structure of work itself. Thus no alternative, "liberatory," science is possible. Habermas writes:

> Technological development . . . follows a logic that corresponds to the structure of purposive-rational action regulated by its own results, which is in fact the structure of *work*. Realizing this, it is impossible to envisage how, as long as the organization of human nature does not change and as long therefore as we have to achieve self-preservation through social labor and with the aid of means that substitute for work, we could renounce technology, more particularly *our* technology, in favor of a qualitatively different one.[33]

To speak of nature as another subject, as a possible (communicative) partner in the creation of a "common universe," Habermas argues, is to use the language of in-

teraction in a realm constituted by work.[34] Marcuse's position thus involves a fundamental category mistake—nature cannot be "dominated" and cannot be "liberated"
either, because concepts such as "domination" and "liberation" are applicable only to
relations between subjects, and the nature that is the object of science and technology
is not a subject. The intrinsic link between the structure of science and the structure of
work means that science is in its essence oriented toward the prediction and control
of nature, not to an oxymoronic "liberating mastery" of it based on a quasi-communi-
cative relation in which it is somehow freed "to become what it would like to be."

The "New Science" thus stands revealed as a romantic dream—to have a technology that is somehow not a technology, to satisfy human needs without controlling
nature. It is the dream of a nature with whom we could speak, of a nature that is itself
a moral agent and with whom a reciprocal moral relation is a possibility. Like all romanticism, it correctly rejects the claims of Science to intellectual hegemony, but incorrectly thinks it can do this only by rejecting Science's validity as such. Habermas
sees the real situation as more complex, and one in which the success of science and
technology (not as a prop for domination, but as a way of satisfying real human needs)
cannot, and need not, be denied. It is the conflation of the technical with the communicative interest, not the technical interest itself, that reinforces domination. If contemporary scientism, in reducing all social questions to technical ones, is guilty of such a
conflation, so too—in the opposite way—is the Marcusean New Science.

"Interests" and Interests

Let me briefly indicate two problems with Habermas's theory of the knowledge-
constitutive interests underlying science.[35] Neither of these have escaped the notice of
commentators, but few have offered an account of their systematic meaning and connection.[36] I want to suggest that they have a common source in the very antinomy that
I began by claiming was endemic to Western Marxist discussions of science, and that
when taken seriously they significantly weaken Habermas's critique of Marcuse.[37]

One problem has to do with the ambiguous status of nature.[38] On the one hand,
as the object of natural scientific inquiry, nature for Habermas must be a *constituted*
realm—constituted in the context of the technical interest in prediction and control of
the environment. This is the Kantian moment in Habermas's "quasi-transcendentalism"—the nature of natural science has no objective reality independent of human
interest. On the other hand and at the same time, nature is conceptualized by Habermas as somehow the *source* of the interests he describes—it is the contingent "natural"
structure of *this* particular species that gives rise to the interests indicated by work and
interaction (not, as in a full-blooded Kantianism, the necessary structures of "any rational being whatsoever"). Thus Habermas is forced to distinguish between a "nature-
in-itself" that gave rise to the human species and a "nature-as-it-appears-to-us" that we
constitute through work and investigate with natural science. And unlike Kant he
needs not simply to assert the existence of this quasi-noumenal realm but in fact to

know something about it—precisely that, and how, it gave rise to the specific interests he claims to be fundamental.

But how is this knowledge possible? The whole thrust of Habermas's epistemology is to insist on the connection of knowledge and interest, but knowledge of "nature-in-itself" would have by definition to be independent of any interest. Thus Habermas has to claim for his own theory exactly the objectivity and freedom from interest that he denies is possible for natural science. The result is an antinomy—the account of knowledge-constitutive interests can get off the ground only by presupposing the possibility of the very kind of knowledge that the account claims cannot exist.

A second problem with Habermas's account is its uncritical acceptance of a positivist view of natural scientific method. It is crucial to Habermas's distinction between the natural and the social sciences that the former are essentially monological—a natural scientist engages in solitary observation of "data," with the intention of developing a "theory" that will both explain the observed "facts" and allow future ones to be predicted. At the level of observation, there is no role for communication with others, or for the (linguistically mediated) interpretation of facts in the light of previous commitment to theories. Natural science, further, is described by Habermas as cumulative, marked by progress, and hypothetico-deductive in its explanatory structure.[39]

Thus although he rejects positivism's epistemology, Habermas remains quite beholden to positivism's account of method. What is curious, of course, is that by the mid-1960s this account had already come under serious attack within the positivist tradition itself.[40] Indeed, as Mary Hesse and others have persuasively argued, the growing "post-empiricism" in philosophy of science of the 1960s had already begun to develop a new account of natural science that made it look remarkably similar to Habermas's account of the "hermeneutic" *social* sciences, emphasizing exactly the constitutive role of interpretation and communication and the consequent difficulty in separating theoretical framework from data that Habermas had taken as characteristic of social science only.[41] Natural science turned out to be just as dialogical, just as discursive, just as dependent on the social, as the *Geisteswissenschaften*.[42]

Habermas's notion of human interests as constitutive of the very objects of knowledge, of course, can be seen as asserting something like the famous "theory-ladenness" of observation post-empiricists such as Kuhn emphasized.[43] In fact there is a crucial difference—whereas for Habermas the "interest" that underlies the facts natural sciences discover is the single, species-wide interest in prediction and control of the environment, for Kuhn (and post-empiricism generally) the more significant point is that facts can be "laden" with *particular* scientific theories, and hence are constituted within *particular* communicative contexts.

Post-empiricism goes further than Habermas, that is, by asserting not only that "interests" (or "paradigms") are constitutive of the "facts" scientists observe, but also that these interests are *historical* interests, changing over time in a process that has much more in common with the circular hermeneutic process of successive reinter-

pretations of a text than with the cumulative and progressive "growth of knowledge."
More recent work in the philosophy of science, particularly that influenced by sociological studies of science, has begun to show in detail the extent to which the "nature"
investigated by natural science is a *constituted* nature, constituted not by some abstract
species-interest in prediction and control, but by a complex set of scientific practices
that instantiate what Kuhn called a "paradigm."[44] The "nature" of natural science in
this sense is a social construct, constituted in and through the interaction not only
among those who investigate it but also within the wider community as well. (I will
return to this point in the next section.)

It is this conclusion that Habermas cannot accept. He wants scientific knowledge
to be interested but not *too* interested—subject to an abstract and species-wide universal interest, but not to any specific social interests. To assert that the object realm
constituted by natural scientific practice reflects real social interests, and changes as
they change, would threaten to unleash a relativism that Habermas finds deeply dangerous. In a remarkable passage from his inaugural lecture at Frankfurt, he writes that
"the glory of the sciences is their unswerving application of their methods without
reflecting on knowledge-constitutive interests," and indeed in the illusion of "disinter-
estedness" and objectivity under which they labor. "False consciousness has a protective function," he goes on; "it was possible for fascism to give birth to the freak of a
national physics and Stalinism to that of a Soviet Marxist genetics . . . only because the
illusion of objectivism was lacking."[45] The admission is telling: although as a theoretical matter Habermas thoroughly rejects the positivist faith in the "objectivity" and
"value-freedom" of science, in practice he shares positivism's fear of the political consequences of the loss of such faith. Any conscious acknowledgment of the irreducibly
"interested" character of science threatens to dissolve the boundaries between science
and politics, and hence to produce the kind of politicization of science that the
Lysenko episode is supposed to caution us against.

But if science is necessarily linked to "interest" those boundaries have already
dissolved, and Lysenkoism *is* a danger. This is the other form of Habermas's antinomy—an "interested" science is inevitably political, and hence the sharp separation
he draws between the technical and the practical interests cannot be maintained. Real
interests are always already social interests, communally interpreted and discursively
(and practically) developed; but if this is so, how could one particular interest (the
technical one) be specified by its independence of the hermeneutic dimension, and
hence of the social? Habermas's practical solution to this problem scarcely seems serious—to encourage the development of false consciousness among scientists, as if
somehow not telling them that their knowledge is linked to interest would prevent the
wrong (fascist, Stalinist) interests from infecting their work. But it sheds light on the
essential ambiguity of his theoretical solution: to employ a notion of interest that somehow is *not* social—a universal "interest of the species," built into its very structure, and
so arising prior to any society or history.

But to call an interest universal is not really to make it any less of an interest, nor to break the link between it and the social. Scientific knowledge is still relativized by such a move, but now to "the species" as a whole rather than to any particular social order. Thus instead of solving the problem of relativism, Habermas's move simply shifts it back in time, where it promptly reappears in another form—exactly as the problem of the status of "nature in itself" discussed above. Instead of wondering how a socially interested science could validly claim to know anything about the world of extra-social nature, we are left to wonder how Habermas's quasi-transcendentalism could claim to know anything about a noumenal nature that gives rise to the technical interest. Having a species-subject rather than a social subject constitute nature may block the politicization of natural science, at least in a narrow sense, but only at the cost of reproducing the Kantian problem of the thing-in-itself. To defend his concept of an "interest" that is not really an interest, Habermas has to posit a "nature" (in itself) that is not really the nature of our experience. This is the deep structure that connects the two problems with Habermas's account of science I have sketched out.

He thus ends up on the reefs of the antinomy that has bedeviled Western Marxism from the beginning. On the one hand his debt to German idealism leads him to reject positivist claims for the "objective" character of natural science, and so to acknowledge that the nature it investigates is a constituted realm, but on the other his commitment to materialism means the validity of natural science must be carefully protected from any hint of relativization to the social. The only way to do this, as we have seen, is by a theory of knowledge as constituted by "interests" that are not the interests of any real subject, and that have their origin in a noumenal nature whose relation to the empirical nature we know remains entirely mysterious. The position is antinomical on its face; and further, with the recognition by post-empiricism that natural science contains an essentially interpretive dimension, and that the interpreters are real historically and socially situated scientists, it turns out to involve a deep misunderstanding of scientific practice.

Nature as a Social Construct

But then Habermas's critique of Marcuse collapses. For what was essential to this critique was the drawing of an ontological distinction between the natural and the social—a distinction that itself depended on the methodological one between the natural sciences and the *Geisteswissenschaften*, and between social interests and the pre-social, universal ones operative in science; but in the context of post-empiricist accounts of science none of these distinctions survive. Habermas is right to assert the interested character of knowledge, wrong to deny that this means that all knowledge, including natural scientific knowledge, is intrinsically connected to the social.

In this sense it is Marcuse who gets the best of the argument. Science *is* always a historical project, and *does* "project a universe"; and thus it *does* make sense to talk of the connection between the "value-free" universe projected by contemporary science and the alienating social order in which that science is embedded, and to raise the possibility of an alternative science that might "arrive at essentially different concepts of nature and establish essentially different facts." A new society based on a new sensibility could then plausibly be expected to bring along with it such a new science, and a new technology—and, yes, a new nature as well.

"Nature is a historical entity," Marcuse writes;[46] post-empiricists have shown us how this kind of claim is to be understood. Their crucial discovery has been that the very objects science takes itself as describing seem to change as science does, and thus that such objects must be seen as constituted by the practices that make up the (communicatively produced and reproduced) "paradigm" in which scientists are at any given moment working. From this follows the impossibility of any appeal to One Unchanging Nature underlying all our different scientific accounts, and hence the impossibility of rating theories as to their closeness to a "correct" description of "nature in itself." In this sense, as I have suggested, the nature of natural science can be described only as socially constructed; science gives us no magical access to a noumenal nature independent of the socially constituted one that is the object of its practices.

The "society" constituting nature cannot be limited to the scientific community alone. As "externalist" accounts of the history of science have always emphasized, scientists are simultaneously part of the broader social order, and hence the categories and assumptions they bring to their scientific work, and to their view of nature, will inevitably owe much to those current in their cultural environment.[47] This works both ways—for, as Bruno Latour and others have recently emphasized, over time science itself has become an influential social power, radically transforming not only the social worldview but quite literally the social world by rebuilding it to allow the more efficient employment of laboratory techniques.[48]

Of course "Science" is only the latest of social projects and social institutions to remake the world in this way. Our relation to the world is always an active, transformative one, in which we work to remake the world in accordance with social ideas and social goals. It is thus not just the nature of natural science that is socially constructed, but "nature" in the sense of the everyday environing world we inhabit as well.[49] That different societies have different "views of nature" is a commonplace of historical and sociological research; but there is more to it than that. In our social practices we produce and reproduce a world more or less adequate to our "view" of it, and in this sense it is quite literally a social construct.

The world we inhabit—the increasingly urbanized world in which the existence of things like electricity, aviation, telephones, and so forth is taken for granted—is a *constructed* world, the product of human labor; and even the world we explicitly call "natural"—the Niagara Falls whose flow is regulated for the pleasure of tourists, the Yellowstone whose fires are extinguished or allowed to burn in accordance with current social attitudes—is separated by no clear distinction from the "artificial" one.[50]

The world that surrounds humans is always already a transformed world, always already the effect of previous human action; this is clearest of course in the case of the rapid and spectacular world-changing brought about by science and technology, but in fact has held from the very beginning, implicit as it is in the structure of what Habermas calls "work."

My point is thus that "nature is a social construct" not only in the sense that what counts as "natural" and how it is experienced are socially variable, but also in two other senses—first, in the sense that the very nature of natural science is "constructed" by the community of scientists through their practices (both narrowly, in the laboratory setting, and more broadly through their own participation in wider social processes), and second, in the more literal sense that the environment we naturally inhabit is always itself the result of prior social action. At no level do we have access to some "nature in itself" independent of human activity; nature is rather always something we constitute through our social practices—through our "work" (which is inseparable from the social structures that organize it) and through our "interaction" (which is inseparable from the processes of material production and reproduction that condition it). There is neither any sense in nor any need for the idea of a noumenal nature "underlying" this socially constructed one.

But then a change in our social practices would mean a change in the world. This is just the point we saw Marcuse making earlier, in his talk of a "new sensibility" that if successful would "literally transform . . . reality," and of the task of the "new science" as the "reconstruction of reality" or "the transformation of nature."[51] "Nature is a historical entity"; this is why the notion of an alternative natural science discovering essentially different facts is not an impossibility. It is just this conclusion, with its implications of a politicized science and technology, that worries Habermas. His quasi-transcendentalist strategies are an attempt to hold on to the constituted character of nature without granting its historical character; but as we have seen they only issue in antinomies. It is finally the active, historical—that is, social—character of our relation to nature that Habermas cannot acknowledge, and that Marcuse seems to assert. If a new social order would produce for itself a new worldview, this simply means—it would inhabit a new world.[52]

Marcuse's Naturalism

And yet: to recognize this, which I think is the key to understanding the sense in which Marcuse is right about the "new science," is also to begin to recognize the sense in which he is quite wrong as well. For in his talk about the new approach to nature a "new science" and "new technology" would instantiate, he constantly speaks of something like a noumenal nature, a nature in itself that previous technology has repressed and that the new one will somehow free. Dominative technology, he writes, "offends against certain objective *qualities* of nature. . . . The emancipation of man involves the recognition of . . . truth in things, in nature."[53] He associates the new sensibility with Platonic anamnesis—the new science would be a form of "*recollection*: . . . the re-discovery of the true *Forms* of things, distorted and denied in the established reality."[54] The liberation of nature would mean that "nature's *own* gratifying forces and qualities are recovered and released,"[55] thus allowing nature for the first time to "exist '*for its own sake.*' "[56] But what is the "it" being discussed here?

There is a contradiction in Marcuse's position that can be quite clearly traced in the chapter on nature in *Counterrevolution and Revolt*. On the one hand, as we have seen, he asserts that "nature is a historical entity" and eloquently insists that the role of a new science and a new technology is to rebuild the world; but on the other hand he constantly writes as though the model for this rebuilt world is to be found somehow in a noumenal nature's "own" "objective" or "inherent" qualities. In an insightful analysis of Marx's *1844 Manuscripts*, Marcuse emphasizes the active character of sensibility and the world-constituting role of human practice and human knowledge; but if knowledge is active and world-creating, then it is not clear how we could ever know anything about what nature "in itself" is like, or what talk about its "objective" qualities would mean.

In fact the reference to Marx turns out to be, in the context of Marcuse's account of the "new sensibility" 's *content*, really a red herring.[57] For it is not the active charac-ter of knowledge that the new science is supposed to emphasize, but rather (and quite inconsistently) its *receptive* character—"the faculty of being 'receptive,' 'passive,' " he writes, "is a precondition of freedom: it is the ability to see things in their own right";[58] earlier he describes the central characteristics of a "non-exploitative" relation to nature as "surrender, 'letting-be,' acceptance."[59] The Heideggerean language is evocative, but is simply inconsistent with the thrust of the argument, which must emphasize that the human relation to nature is in its very essence active and transformative, and thus that no "receptive"—and entirely ahistorical—acceptance of "nature in itself" can even be coherently imagined.

The theme of a call for such "receptivity" runs deep in Marcuse's work.[60] In *Eros and Civilization* "productiveness" is explicitly associated with toil, and with the per-formance principle that his utopianism wants to overcome. It is "receptiveness" that is connected with "joy," with "play," with the "absence of repression," and so with the pleasure principle which the advance of civilization was forced to renounce but which the development of technology makes it possible once again to satisfy.[61] Activity al-most always appears in that work as negative, as "domination"; freedom is associated with fulfillment, with completion, with an end to striving.[62] Rejecting the image of Prometheus ("the culture-hero of toil, productivity, and progress through repression")[63] in favor of those of Orpheus and Narcissus, Marcuse writes that the latter

reconcile Eros and Thanatos. They recall the experience of a world that is not to be mastered and controlled but to be liberated—a freedom that will release the powers of Eros now bound in the repressed and petrified forms of man and nature. These powers are conceived not as destruction but as peace, not as terror but as beauty. It is sufficient to enumerate the assembled images in order to circumscribe the dimension to which they are committed: the redemption of pleasure, the halt of time, the absorption of death; silence, sleep, night, paradise . . .

and ends by quoting Baudelaire's "ordre et beauté,/luxe, calme, et volupté."[64] Such talk—linking Eros and Nirvana, pleasure and death, in a way Freud would have found unrecognizable—reveals a deep ambiguity in Marcuse's position, in which the emphasis on the active, transformative role of humans in nature clashes with a romantic yearning for passivity, for "silence" and "sleep," for delicious surrender to the powers of nature.[65] Nature here is not to be transformed, but rather "let be"—the "powers of Eros" are inherent in a nature-in-itself that stands prior to human action, but which human action has repressed. All we need to do is cease our dominative struggle, and a liberated nature will liberate us.

Thus the insight that "nature is a historical entity" is not carried through. If Habermas fails to see that the active character of our relations with nature is inseparable from their social character, Marcuse here fails to see the opposite—that a historical entity is always already the product of human action and hence that "passivity" or "receptivity" or "surrender" to it is not even an option. Instead he constantly tries to appeal to a noumenal (and thus ahistorical) nature independent of human action as the foundation for social critique. The result is a curious tendency toward biologism in his work. The project of *Eros and Civilization* is precisely to ground critical theory in the instincts—in the memory of the instinctual gratification repressed by civilization but preserved in the realm of the unconscious and of imagination. The same theme recurs in *An Essay on Liberation*, where he writes of "a biological foundation for socialism."[66] The assumption seems to be that only a social critique founded in nature can claim validity for itself.

To be sure, in Marcuse's works of the 1960s and 1970s his biologism is also often expressed in a quite different and more surprising claim—that revolutionary change literally requires the production of *new* instincts, even of a new biology for humans, so as to produce "new needs."[67] Contemporary society has generated needs for consumption, for competition, for waste, Marcuse writes, that have "sunk down" into the organic dimension;[68] a liberated society would thus have to replace those needs with new ones—"for peace, . . . for calm, . . . for the beautiful," and so forth.[69] The idea would be to produce "men who have developed an instinctual barrier against cruelty, brutality, ugliness."[70] Here the "historical" character of internal nature is certainly asserted,[71] but only in such a way as to betray the depths of Marcuse's biologism and naturalism—as if the only manner in which revolutionary action could be justified would be if it became a matter of natural impulse. Nature continues here to be the foundation of social change, but perversely we are now supposed to change nature first in order to make it possible for it in turn to change us.

If nature really is historical, though, then "changing nature" is not so much a precondition for social change as it is an effect of it; building a new society does not require humans changing their own instincts, but rather changing their own minds. This illuminates what is absolutely right about Habermas's critique: like the Marxist scientism earlier Western Marxists had rejected, Marcuse wants to make social revolution a requirement of nature, and thus takes what are essentially ethical questions and tries (by fiat) to turn them into biological ones.[72] The result is to remove social critique from the communicative realm and hence from its requirement of discursive justification: for the young rebels Marcuse supports, he says, the struggle "is not a question of choice; the protest and refusal are parts of their metabolism."[73] The rhetoric is impressive, but misguided: where there is no choice, there can be no liberation either.[74] The mistake is to try to ground social theory in nature, which is to say in the extra-social—a move that always ends up making humans subservient to things, and that Marx had criticized as fetishism and Lukács as reification.[75]

But we can see Marcuse's motivation—for where else could social theory be grounded? He too, like Habermas, fears relativism. Without an appeal to some ahistorical nature independent of human will and human choice, it does not seem possible to justify criticism of a social order that admittedly "delivers the goods" and apparently satisfies most people's needs, or even to give a content to the critique. The solution seems to be to posit a nature-in-itself whose *own* "needs" are being violated by such a system, and then to make the revolution in its name. But Marcuse's own argument, as we have seen, shows that no such nature really exists, since "nature" is always a historical entity, always already the product of social construction. And to speak of "constructing" nature in such a way as to "let it be," of allowing it to exist "for its own sake" or to realize its inherent telos, will not work—for again there is no "it" there, no model for the "truly natural" separate from what we *do*.

But then Marcuse's problem turns out to be identical to Habermas's, and to stem from the same source. Both writers follow lines of thought that lead to the conclusion that the external world of nature is actively constituted through social practices; both try to avoid the apparently relativist implications of this conclusion by attempting to ground those practices (from the beginning of the species, for Habermas; "after the revolution," for Marcuse) in a noumenal—that is to say, un-constituted—nature. But neither is able to explain how we can have access to this noumenal nature, given the original assertions about the active character of our relation to nature. The problem here is the one I began by claiming was endemic to Western Marxism from the start—how to reconcile an account of knowledge as active and social on the one hand with the "materialist" commitment to a nature independent of the human on the other.

Yet if the recognition of the active character of knowledge, and the constituted character of nature, reveals the shortcomings of both sides of the dispute between

Habermas and Marcuse about technology, it at the same time shows there to be something right about each side as well. Habermas's emphasis on the role of human interest in our knowledge of nature shows that nature is *not* itself a subject, rather being the object of our constitutive acts. We cannot have a communicative relation to it—it doesn't have "rights," it can't be "liberated," it does not "speak to us" or "await the revolution." It is through and through a historical product, the result of human action. For just this reason, however, something like Marcuse's New Science *is* possible: a new social order would mean a new mode of action, and would bring with it a new world (and even—as post-empiricism makes clear—"new facts").

Marcuse's error is to want to ground this new social order in some ahistorical "objective qualities" of nature in itself, which is just what the very logic of the argument makes impossible. In doing so he regresses behind the insights of classical German idealism: nature is taken as independent of us, as existing "in itself" prior to and separate from any human constitutive acts—in Habermas's language, independently of any interest—and the correct role for human action is to follow it and to "let it be." In *Counterrevolution and Revolt* Marcuse writes of the Hegelian notion of a consciousness that "discovers that *we* ourselves are behind the curtain of appearance"—that the objective world it took as independent and external is in fact the product of its own socially mediated actions.[76] But it is just this "we" that is missing in his talk of inherent qualities, truths of nature, basic instincts—the "we" who make the history within which nature always appears.

If we grant that nature is historical, and socially constituted, though, the problem of relativism does seem to beckon—how can one society's approach to nature (internal and external) be critically compared with any other's? Hegel knew the solution to this problem—which is the problem of how to ground a critical theory of society in something other than nature: there is a difference between the acts of a world-constituting subject not yet aware of what it is doing and those of a subject who has self-consciously grasped its responsibility for the world it produces. Habermas, we saw, fears the consequences of a self-reflective science, one that knows its own interested character; but it is just this, it seems to me, that is the best candidate for the Marcusean New Science. What is wrong with the society we live in is precisely that in it human beings' own responsibility for the environment they inhabit (the "social" environment, in the first place, but the so-called natural one as well) is systematically hidden from them—this is the phenomenon Marx called alienation. A non-alienated society would be one where that illusion is punctured, and humans consciously and explicitly assert their responsibility for the world, transforming it on the basis of needs that are discursively expressed and social decisions that are democratically made. The science and the technology *that* society would have would have to be radically different than the one we know today. No longer in the thrall of positivist myths about the possibility of a pure description of an external reality (myths that themselves simply reflect the prevailing reification), such a science would finally know itself to be social, to be historical, to be "interested," and hence would know its own connection to the world it helps create. It is that very knowledge that breaks the link between science and domination Western Marxism has always feared; without the seductive faith in its own "neutrality" science would know itself as an instrument of liberation—not silent nature's, but our own.

Notes and References

1. Georg Lukács, *History and Class Consciousness* (Cambridge, MA: MIT Press, 1971), p. 234.

2. Ibid., p. 10.

3. Marcuse's position was expressed in some detail in his *One-Dimensional Man* (Boston, MA: Beacon Press, 1964), and then further developed in: "Industrialization and Capitalism in the Work of Max Weber," in *Negations* (Boston, MA: Beacon Press, 1968); "The End of Utopia," in *Five Lectures* (Boston, MA: Beacon Press, 1970); *An Essay on Liberation* (Boston, MA: Beacon Press, 1969); and *Counterrevolution and Revolt* (Boston, MA: Beacon Press, 1972), esp. chapter 2. Habermas's critique is found in the collection originally published in Germany as *Technik und Wissenschaft als "Ideologie"* (Frankfurt: Suhrkamp, 1968); the title essay and several others appear in English in *Toward a Rational Society* (Boston, MA: Beacon Press, 1970), while another appears in *Theory and Practice* (Boston, MA: Beacon Press, 1973). See also his *Knowledge and Human Interests* (Boston, MA: Beacon Press, 1968) for an account of his overall epistemological position at the time.

4. For discussion of the dispute and its wider implications within the history of Western Marxism, see C. Fred Alford, *Science and the Revenge of Nature* (Tampa, FL: University Presses of Florida, 1985); Thomas McCarthy, *The Critical Theory of Jürgen Habermas* (Cambridge, MA: MIT Press, 1978), pp. 16–40; Paul Connerton, *The Tragedy of Enlightenment* (Cambridge, MA: Cambridge University Press, 1980), chaps. 6–7; Norman Stockman, "Habermas, Marcuse, and the *Aufhebung* of Science and Technology," *Philosophy of Social Science* 8 (1978): 15–35; Ben Agger, "Marcuse and Habermas on New Science," *Polity* 9 (Winter 1976): 158–81; Stanley Aronowitz, *Science as Power* (Minneapolis, MN: University of Minnesota Press, 1988), pp. 159–68.

5. Max Horkheimer and Theodor Adorno, *Dialectic of Enlightenment* (New York: Seabury Press, 1979).

6. For example, see *Eros and Civilization* (New York: Vintage Books, 1962), pp. 136–43.

7. Ibid., p. 142; cf. also p. 95. See Alford, *Science and the Revenge of Nature*, pp. 37–43.

8. For discussions of Marcuse's ambivalent relation to technology, see especially William Leiss, *The Domination of Nature* (Boston, MA: Beacon Press, 1974), pp. 207–12; Alford, *Science and the Revenge of Nature*, chaps. 3–4; Andrew Feenberg, "The Bias of Technology," in *Marcuse: Critical Theory and the Promise of Utopia*, edited by R. Pippen, A. Feenberg, C. P. Webel (South Hadley: Bergin and Garvey, 1988), pp. 225–56; Douglas Kellner, *Herbert Marcuse and the Crisis of Marxism* (Berkeley, CA: University of California Press, 1984), pp. 323–29.

9. Thus in *Eros and Civilization*, alongside the talk of "the inevitably repressive workworld" (p. 178) and the assertion that it is only "the sphere outside labor which defines freedom and fulfillment," one can also find Marcuse speaking very differently—for example, of "a change in the character of work by which the latter would be assimilated to play" and so "would tend to become gratifying in itself without losing its *work* content" (pp. 195–96).

10. See "The End of Utopia," pp. 62–63.

11. "Industrialization and Capitalism in the Work of Max Weber," pp. 225, 224.

12. For useful discussion of Marcuse's later views, see Alford, *Science and the Revenge of Nature*, especially chap. 4, as well as Leiss, *The Domination of Nature*, Appendix; and Kellner, *Herbert Marcuse and the Crisis of Marxism*, pp. 330–38.

13. See *One-Dimensional Man*, pp. 156–69.

14. Ibid., pp. 166–67; see also pp. 230–33. "This development," Marcuse writes (p. 233) "confronts science with the unpleasant task of becoming *political*—of recognizing scientific consciousness as political consciousness"; as we will see, it is precisely this kind of self-reflexive recognition that Habermas wants to avoid. See below, p. 31.

15. See *An Essay on Liberation*, pp. vii–ix; *Counterrevolution and Revolt*, pp. 31–33, 74–78. See Kellner, *Herbert Marcuse and the Crisis of Marxism*, chapter 9.

16. *Counterrevolution and Revolt*, p. 60.

17. Ibid., p. 74.

18. Ibid., p. 67.

19. *An Essay on Liberation*, p. 31.

20. See "The End of Utopia," p. 68.

21. *Counterrevolution and Revolt*, pp. 2–3.

22. *One-Dimensional Man*, p. 236. Also see Leiss, *The Domination of Nature*, p. 212.

23. *Counterrevolution and Revolt*, p. 66; the phrase is translated differently in Theodor Adorno, *Aesthetic Theory* (London: Routledge & Kegan Paul, 1984), p. 101. "Paradoxical" might be too weak a word here; I argue later that there is something finally incoherent about the notion of a "transformation" of nature that nonetheless reveals its "inherent" qualities.

24. *Eros and Civilization*, p. 151.

25. *An Essay on Liberation*, pp. 45, 31. Cf. *Eros and Civilization*, p. 173: "the play impulse would literally transform the reality."

26. See Alford, *Science and the Revenge of Nature*, pp. 43–45.

27. The fullest statement of this theory is in *Knowledge and Human Interests*. Habermas no longer holds to the theory in this particular form; see note 37 below. For useful critical accounts of Habermas's position (both earlier and later) see McCarthy, *The Critical Theory of Jürgen Habermas*; David Held, *Introduction to Critical Theory* (Berkeley, CA: University of California Press, 1980), part two; Seyla Benhabib, *Critique, Norm, and Utopia* (New York: Columbia University Press, 1986); David Ingram, *Habermas and the Dialectic of Reason* (New Haven, CT: Yale University Press, 1987).

28. See *Knowledge and Human Interests*, pp. 308–11. On the notion of the "quasi-transcendental," see ibid., pp. 194–95.

29. See *Toward a Rational Society*, pp. 91–93.

30. See *Knowledge and Human Interests*, chap. 9. Habermas also asserts the existence of a third, "emancipatory," interest, of less importance for this discussion.

31. See ibid., pp. 304–308.

32. *Toward a Rational Society*, p. 88.

33. Ibid., p. 87.

34. Ibid., p. 88.

35. This section summarizes some arguments from my "Habermas and Science," *Praxis International* 8 (October 1988): 329–49, esp. sections III, IV, and VI.

36. See McCarthy, *The Critical Theory of Jürgen Habermas*, chap. 2; Alford, *Science and the Revenge of Nature*, chaps. 5–6; Held, *Introduction to Critical Theory*, pp. 389–98; Richard Bernstein, *The Restructuring of Social and Political Theory* (Philadelphia, PA: University of Pennsylvania Press, 1976), pp. 219–25, and "Introduction" to *Habermas and Modernity*, edited by R. Bernstein, (Cambridge, MA: MIT Press, 1985), pp. 8–15.

37. It is worth noting that Habermas himself has given up the views under discussion here, and has on several occasions acknowledged their inadequacies. See especially the "Postscript to *Knowledge and Human Interests*," *Philosophy of Social Science* 3 (1973): 157–89, and *Theory and Practice*, pp. 14ff. But in his later work questions of natural scientific method and its justification no longer form a major topic of concern; and to the extent they do, his account continues to suffer from essentially the same problems. See Vogel, "Habermas and Science," pp. 338–42. (In a 1984 inter-

view Habermas is quoted as saying that "I still consider the outlines of the argument developed in the book [*Knowledge and Human Interests*] to be correct"—see *Habermas: Autonomy and Solidarity,* edited by P. Dews, London: Verso, 1986, p. 152.)

38. See McCarthy, *The Critical Theory of Jürgen Habermas,* pp. 110–25. See also Joel Whitebook, "The Problem of Nature in Habermas," *Telos* 40 (Summer 1979): 41–69, and Henning Ottmann, "Cognitive Interests and Self-Reflection," in *Habermas: Critical Debates,* edited by J. B. Thompson and D. Held (Cambridge, MA: MIT Press, 1982), pp. 79–97, as well as Habermas's own discussion of the problem at pp. 238–50 of the same volume. Whitebook and Ottmann offer what might be called a "Marcusean" critique of Habermas's view of nature; as will be clear below, my own critique in fact runs in a quite different direction.

39. See *Knowledge and Human Interests,* pp. 191–93; also p. 161, pp. 308–309, and chapter 6 in general.

40. Habermas has conceded that "in the light of the debate set off by Kuhn and Feyerabend, I see that I did in fact place too much confidence in the empiricist theory of science in *Knowledge and Human Interests.*" "A Reply to My Critics," in *Habermas: Critical Debates,* edited by Thompson and Held, p. 274.

41. See Mary Hesse, *Revolutions and Reconstructions in the Philosophy of Science* (Bloomington: Indiana University Press), pp. 169–73. See also Alford, *Science and the Revenge of Nature,* pp. 77–78; Joseph Rouse, *Knowledge and Power* (Ithaca, NY: Cornell University Press, 1987), pp. 195–96; Karin D. Knorr-Cetina, *The Manufacture of Knowledge* (Oxford: Pergamon Press, 1971), pp. 143ff.

42. To be sure, Habermas had insisted—*against* positivism—on a crucial role for discourse among the community of scientists, and hence for hermeneutics, in the processes of theory-choice and theory-justification. But for him this served only as an argument against the positivist identification of "monological" natural scientific knowledge with knowledge as such, never as an argument against the positivist account of natural scientific knowledge itself. See *Knowledge and Human Interests,* p. 137.

43. See, for instance, his critique of Popper in "A Positivistically Bisected Rationalism," in *The Positivist Dispute in German Sociology* (London: Heinemann, 1976), pp. 200–209.

44. For example, see Barry Barnes, *Scientific Knowledge and Sociological Theory* (London: Routledge & Kegan Paul, 1974); David Bloor, *Knowledge and Social Imagery* (London: Routledge & Kegan Paul, 1976); Bruno Latour and Steve Woolgar, *Laboratory Life,* 2nd ed. (Princeton, NJ: Princeton University Press, 1986); Knorr-Cetina, *The Manufacture of Knowledge;* Bruno Latour, *Science in Action* (Cambridge, MA: Harvard University Press, 1987). See also Rouse, *Knowledge and Power,* esp. chapter 6.

45. *Knowledge and Human Interests,* p. 315.

46. *Counterrevolution and Revolt,* p. 59.

47. Developing this insight beyond the simplistic form it has traditionally taken, especially in Marxist discussions of science from Engels on, has been a central theme in recent work in the philosophy of science, particularly in the "ethnographic" work that began with Latour and Woolgar's *Laboratory Life;* see, for example, the essays collected in *Science Observed,* edited by K. Knorr-Cetina and M. Mulkay (London: Sage, 1983). The question of the social rootedness of science, and more particularly of its roots in gender politics, has received a great deal of useful discussion within feminist theory. See for example, Carolyn Merchant, *The Death of Nature* (San Francisco: Harper & Row, 1983); Evelyn Fox Keller, *Reflections on Gender and Science* (New Haven, CT: Yale University Press, 1985); Sandra Harding, *The Science Question in Feminism* (Ithaca, NY: Cornell University Press, 1986). On the vicissitudes of "externalist" views, and Marxist philosophy of science in general, see Aronowitz, *Science as Power.*

48. See Bruno Latour, "Give Me a Laboratory and I will Raise the World," in *Science Observed,* edited by Knorr-Cetina and Mulkay, pp. 141–70. See also Rouse, *Knowledge and Power,* esp. chapters 4 and 7.

49. Latour's point is that as science develops these two cease to be distinguishable in any case. See "Give Me a Laboratory and I will Raise the World," pp. 153–56.

50. On Niagara Falls, see George A. Seibel, *Ontario's Niagara Parks* (Ontario: Niagara Parks

Commission, 1987), pp. 166–73, and Martin H. Krieger, "What's Wrong with Plastic Trees?" *Science* 179 (February 2, 1973): 447–48. On Yellowstone, see Alston Chase, *Playing God in Yellowstone* (San Diego: Harcourt, Brace, Jovanovich, 1987).

51. See notes 19 and 25. The last quotation is from *Counterrevolution and Revolt*, p. 64.

52. Compare Kuhn's notorious assertion (in *The Structure of Scientific Revolutions* [Chicago: University of Chicago Press, 1962], pp. 121, 150) that after a scientific revolution scientists live in a new world; the point—and, I admit, the problems it raises—is the same.

53. *Counterrevolution and Revolt*, p. 69.

54. Ibid., pp. 69–70.

55. Ibid., p. 67.

56. Ibid., p. 62.

57. Thus by the end of the discussion of Marx, Marcuse concludes that the former's activist view "retains something of the hubris of domination," and needs to be modified (*Counterrevolution and Revolt*, pp. 68–69). The "modification," it seems to me, in fact vitiates the whole argument. See Morton Schoolman, *The Imaginary Witness* (New York: New York University Press, 1984), pp. 125–27.

58. *Counterrevolution and Revolt*, p. 74.

59. Ibid., p. 69.

60. See Schoolman, *The Imaginary Witness*, chapter 3, and pp. 284–86.

61. *Eros and Civilization*, p. 12.

62. See ibid., pp. 99–107.

63. Ibid., p. 146.

64. Ibid., p. 149.

65. On Eros and Nirvana, see, for example, *Eros and Civilization*, pp. 214–15, or p. 247. See also "Freedom and Freud's Theory of Instincts," in *Five Lectures*, pp. 7–8. Cf. Schoolman, *The Imaginary Witness*, pp. 112–16.

66. *An Essay on Liberation*, p. 7.

67. See ibid., pp. 16–17; also *Counterrevolution and Revolt*, pp. 16–17, and "The End of Utopia," pp. 71–72. "What do we need a revolution for if we're not going to get a new human being?" asked Marcuse in a 1977 conversation with Habermas; see "Theory and Politics," *Telos* 38 (Winter 1978–79): 133.

68. *An Essay on Liberation*, pp. 10–11.

69. "The End of Utopia," p. 67.

70. *An Essay on Liberation*, p. 21.

71. For example, see "The End of Utopia," p. 65.

72. See Habermas, "Psychic Thermidor and the Rebirth of Rebellious Subjectivity," in *Habermas and Modernity*, edited by Bernstein, pp. 74–77.

73. *An Essay on Liberation*, p. 63.

74. Thus the last paragraph of *An Essay on Liberation* begins by calling once again for "new instincts" replacing the old competitive ones as the basis for the new society, but ends by quoting "a young black girl" who speaks of the realm of freedom as one where "for the first time in our life, we shall be free to think about what we are going to do" (p. 91); there is no recognition that acting from instinct and "thinking what we are going to do" are *mutually exclusive*.

75. I would suggest that what might be called the *anti-democratic* aspects of Marcuse's thought—the talk of false needs, of repressive tolerance, of the "limits of democratic persuasion and evolution" (*An Essay on Liberation*, p. 17)—have their source in just this mistake. This is very clear in the 1979 discussion between Habermas and Marcuse—see "Theory and Politics," pp. 133–38.

76. *Counterrevolution and Revolt*, pp. 72–73.

3

On the Notion of Technology as Ideology

Robert B. Pippin

Technology as a Political Problem

I T IS AN undeniable fact that a central feature in the history of Western modernization has been an ever-increasing reliance on technology in the production of goods, in services, information processing, communication, education, health care, and public administration. This reliance was anticipated and enthusiastically embraced by the early founders of modernity (Bacon and Descartes, especially), and finally (much later than they would have predicted) became a reality in the latter half of the nineteenth century. Moreover, increasing technological power proved an especially valuable asset in liberal democratic societies. The great surplus wealth made possible by such power appeared to allow a more egalitarian society, even if great inequalities persisted; representatives of such technical power could exhibit, publicly demonstrate, and so justify their power in ways more compatible with democratic notions of accountability; and a growing belief in the "system" of production and distribution as itself the possible object of technical expertise seemed to make possible the promise of a great collective benefit, given proper "management," arising from the individual pursuit of self-interest promoted by market economies.

Since that time such an increasing dependence on technology has been perceived to create a number of straightforwardly political problems and publicly recognized controversies. Commentators came to see that this reliance also came with certain social costs, created difficult ethical problems, and began to alter the general framework within which political discussions took place. Such problems included the following:

1. A greater and greater *concentration of a new sort of social power* in fewer and fewer hands. At least within democratic societies, such a concentration of power might easily become inconsistent with the ideal of democratic control of socially relevant decision making. As noted, while there are deep compatibilities between democratic values and such scientific canons as the public demonstrability of knowledge claims and the public benefits of the ends to which technology can be employed (e.g., public health, agricultural planning, communication), it is also true that the rise of expert

elites posed a certain sort of threat. With the growing sophistication of science and technology, and the difficulties encountered by a lay public in understanding evidence, demonstrations, and the ambiguities and risks inherent in the pursuit of any end, such elites grow progressively less accountable in traditional ways for the exercise of their power, shielded as they are by the claim to greater technical competence.

2. A simultaneous and connected *de-skilling* of the labor force through automation, and more rigid, hierarchical forms of technically efficient administration. In such cases the imperatives generated by competition can promote an increased acceptance of technically efficient monitoring techniques (e.g., typists on centralized computers, whose backspace or delete key usage is closely monitored, and who thus can be held accountable not only for what they do but what they could have done, or the zealous monitoring that the phone company exercises over its operators), job simplification, greater risks to worker safety in order to conform to more efficient machines, and a variety of organizational strategies, all relatively inconsistent with basic, post-Enlightenment ideals of self-respect, dignity, and autonomy.

3. A connected and much noted phenomenon in writers such as Arendt and Habermas: a narrowing of acceptable topics for "public debate," thanks to a greater emphasis on policy issues as technical issues.[1] This amounts to the *"depoliticizing" of public life*, such that much political debate becomes merely a war among competing experts, or an exercise in the manipulation of symbols, a wholly theatrical celebration of rival images and icons, all rather than a collective and substantive deliberation about a common societal direction.

4. A simple increase in the *extent of administrative power* over aspects of daily life. Foucault's claims for micro- and bio-power are relevant here, and the simply massive character of the kind of power made possible by data storage in medicine, government, banking, insurance, etc.[2] An example worrisome to many recently has been the project to map the human genome. Armed with this information and new diagnostic techniques, eventually it may be possible for, say, insurance companies or potential employers to predict from a simple blood test, with some reliability, the chances that an individual will get a stroke or coronary disease, whether he smokes or drinks too much, or even, perhaps, eventually, whether he suffers too much stress, or is too neurotic or too unsociable, etc.

5. An extraordinary new role for science and technology in *national security issues*, requiring diversion of vast resources to ever more expensive weapons research, a diversion that has seriously and perhaps permanently derailed hopes for welfare state capitalism.

6. Cultural complaints that the "technological tail" was beginning to wag the "human dog"; that too many areas of daily life were being modified to meet *the needs of technical efficiency*. Complaints, for example, about being reduced to a number, having to talk to answering or voice-mail machines instead of people, excessively techno-

logical and so "dehumanized" environments for birth, illness, and death, or complaints about the medicalization of mental health issues.

Technology as an Ideological Problem

Often such topics are discussed within some sort of cost-benefit framework and under the assumption of a kind of technological fatalism that the clear efficiency of a reliance on technology makes such continuing or ever-growing reliance more or less inevitable, or at least unproblematically rational. Amelioration of the social costs, and an exploration of options with respect to ethical dilemmas, could, under such assumptions, occur only marginally, as a kind of moral hope, and only after the technological imperative had been basically satisfied (a situation especially obvious in a climate of worry about international competitiveness).

A more radical critique can be detected in those who understand technology itself as a kind of *ideology*. This notion is both a complex and vague category, and its usefulness has suffered a great deal from an increasing use of the term to mean simply a philosophy or belief. But, as a critical concept, the notion was made possible by the Kantian revolution in philosophy and its central claim that there could be "forms" or "conditions" of experience not themselves derived from experience, but "constitutive" of the very possibility of experience. It was this notion of a priori constraints on empirical experience or, more broadly, belief formation that set the stage for Hegel's historicization of these categories, Marx's social theory, and Lukács's use of the notion of "reification" in a full-blown "ideology critique."[3]

Within this tradition, a form of consciousness, or a general, comprehensive categorization of experience, can be ideological in any number of senses.[4] Ideological claims can be claims about the nature of reality, the significance of a social practice, the origin and legitimacy of an institution, the authority of a moral code, or many other sorts of things. To claim that any such general, fundamental orientation to the world is ideological means not only that some inter-connected set of propositions about nature, others, or the cosmos is false, unsupported by evidence or argument, unproven or irrationally believed, but that such an orientation or form of consciousness somehow *prevents*, renders unnoticeable even, contrary evidence or argument. Consciousness itself, the way we originally take up and make sense of things, can be "false." (Of course, this resistance to criticism is not something constructed consciously and strategically, and so the question of whether there could be, or why there should be, this *sort* of blindness, rather than just mistaken, overly optimistic, or ethically inappropriate worldviews, is an important one for *Ideologiekritik*.)

It is controversial whether there are or ever have been "forms of consciousness" with these characteristics, but the notion, especially when applied to some sorts of religious or moral views, is, *prima facie*, plausible, and has also played a major role in deflationary critiques of technology (or technological rationality, or technological "promise"). (Nineteenth-century accounts of religion are probably the most familiar version of this sort of critique: religious practices cannot be successfully explained by reducing religious propositions to unsupported empirical claims, motivated by compelling psychological states. Such practices could come to have the significance they do only within a certain context, one wherein human power over nature is required but unavailable, producing a projection of such aspirations onto a human-like agent whose vast powers can be appealed to in support of human causes. The increase in real power thus helps explain the diminishing significance of religion in modernity, and so forth.) For the technology issue the question thus is: has our "relation to objects" been so influenced by technical instruments, the power of manipulation and production, etc., that our basic sense of the natural world has changed and changed so fundamentally that our reflective ability to assess and challenge such a change is threatened? Has our understanding of others and of social and political life become so shaped by technical imperatives in production, consumption, social organization, daily life, and politics that fundamental possibilities for social existence are seen only (in a "distorted," narrow sense) in terms of such technical imperatives?

Thus, in general, "ideology critics" are more interested in what is *undiscussed* in the modern experience of technology, what an extensive reliance on technology, which often is presented as a value-neutral tool, itself already hides, distorts, renders impossible to discuss as an option. To see technology as an ideology is to see an extensive social reliance on technology and the extensive "mediating" influence of technology in daily life as already embodying some sort of "false consciousness"; again, a way of looking at things not characterizable as simply a matter of false or problematic or narrow beliefs. And this means: such reliance reaches a point where what ought to be understood as contingent, an option among others, open to political discussion, is instead falsely understood as necessary (i.e., the relevant options are not rejected; they are not noted as credible options; hence the "false consciousness"); what serves particular interests is seen, without reflection, as of universal interest; what is a contingent, historical experience is regarded as natural; what ought to be a part is experienced as the whole, and so on.

The Classical Positions and the Classical Problems

This is all a large and much discussed issue, but, in order to make the point I am interested in, I shall need to survey the general terrain from a fairly high altitude, and very quickly. I want first to set out briefly some typical sorts of claims that "technology is ideology," and some of the problems generated by such claims, before first introducing a general objection, and then focusing on one of the most influential recent arguments, Habermas's.

I begin with the most complex view, and one which does not use, and would deliberately avoid, all the notions of ideology critique, Heidegger's. Famously, Heidegger claims that technology embodies an "orientation to Being," and is "ideological" in the sense that, in such an orientation, Being is "forgotten." He claims that modernity itself is "consummated" or "completed" by a technological "en-framing" (*Ge-stell*), that technology exemplifies an understanding of Being, an absolutely fundamental orientation, which completes modern subjectivism and thoughtlessness.[5] (Technology is a "worldview," or pre-theoretical "horizoning" of experience, a view also roughly maintained by Ellul.)[6]

Several commentators have objected to Heidegger's explanation of the "predatory" stance of the modern subject by appeal to an obscure "history of Being," in which, it appears, Being itself is responsible for its own obscuring or for our forgetting Being. Others dispute the way he dates modernity (as originating in Plato), or complain about the ambiguous practical consequences of his critique of modernity. My own view is that the central problem with Heidegger's approach is that the way it addresses the basic historical questions at issue is undialectical and even a bit moralistic. For him, the appeal of the modern emphasis on power, control, and the priority of the self-defining subject seems to be due to a kind of human *hubris*, a self-assertion that often sounds more like a theological account of the Fall than a historical explanation. The possibility that the basic ontological dimension embodied in a technological world-view (the subject-object split) could have been *provoked* historically, or was required in some sense, given the unavoidable and genuine deficiencies and dead ends created by the premodern tradition (understood in its own rather than later terms), is not considered by him.

To be sure, he has his own, infinitely complicated reasons for this neglect, having to do with his own understanding of the History of Being, and how such a history and his (Heidegger's) own role in the destruction of Western metaphysics play roles in the origin of modernity. But I simply note that his own position requires a historical narrative that has been, rightly, I think, the subject of much attack. (I have attempted a fuller assessment of Heidegger's position elsewhere.)[7]

I should also note that Heidegger's approach also ought to remind us of very speculative claims much discussed recently: that the modern fixation on technological power is not uniquely modern or a distortion of anything, but some sort of culmination of the deep connection between *all* knowledge and "the will to power" or a final revelation of the nature of power/knowledge (*pouvoir/savoir*).[8] These Nietzschean and Foucaultean themes, while continuous with Heidegger's approach, in some ways go much farther than Heidegger, who always seems to want to preserve a contrast between the modern "age of the world picture" and some possible alternative. These approaches also raise their own famous questions about what sort of critique or critical knowledge is possible under such assumptions, but that would introduce in this context a major digression.

To return to positions more traditionally identified within the "ideology critique" tradition, we should recall the original Marxist argument that the capitalist claim for its own "rationality," understood as technical efficiency, is ideological.

Marx focuses on the organization of production under liberal or so-called free market capitalism. What are asserted to be the imperatives of technical efficiency are "functionally" ideological in Geuss's sense. The claim for the efficiency of the capitalist organization of technology is only temporarily true. When maintained beyond the early phase of capitalism, such claims become a "socially necessary illusion," functioning to mask social contradictions, actually to impede the development of the forces of production and even greater technical efficiency, and to stabilize and sustain forces of domination that could not be sustained without such illusions.[9]

This position has been most often criticized for its historical limitation to the liberal phase of capitalism. With the onset of state intervention in and management of the economy, ideology critique could no longer be a critique of political economy alone. (There was no longer any such thing as "the economy" operating under its own laws. Or the model of a "base" supporting and, by virtue of its autonomous development, straining against, then being constrained by, superstructure has been eclipsed with the arrival of welfare or state-interventionist capitalism.)

Second, the idea that the growth of productive forces is itself, inherently, emancipatory, or helps expose the ideological character of the justification for historically outmoded relations of production, has been rendered obsolete. The main contribution to growth in the forces of production now comes *from* science and technology, designed and implemented *by* managers and bureaucrats. Adherence to the imperatives of technical efficiency now helps legitimate the entire self-regulating social system.

Third, this notion of system now seems more relevant to social analysis than traditional class-conflict notions. Genuine, clear-cut oppositions between class interests are now rarer. Allegiances are secured through a complex and efficient system of rewards and leisure time, and all perceive themselves to be helping to operate a system of benefit to all, rather than serving the interests of a discrete, identifiable group of others. This means, for someone like Habermas, that reflection can identify an emancipatory interest distorted or repressed in such a system, only as a *species interest,* as an interest of humanity, denied or regulated by such systematic imperatives. (Again, he wants to replace the class conflict/forces-relations of production model with what he calls a "work-interaction" model.)

Finally, the basic charge of suboptimization can be met on its own terms in later capitalism. State capitalism can easily claim to have solved that problem and to have regulated the cycles of early capitalism far more efficiently than other available historical models (certainly better than command economies).

In a radical extension of the scope-of-ideology critique, Horkheimer and Adorno connect the problem of technology to the "dialectic of enlightenment" in modernity in general, and so connect the mastery of nature to the mastery of others, and an attempted legitimation of domination, control, and psychological repression which, they maintain, is ultimately self-undermining and de-legitimating.[10] The most well-known form of such a critique (despite many differences) became Marcuse's attack in *One-Dimensional Man*.[11]

In the account given by Horkheimer and Adorno in *The Dialectic of Enlightenment*, the prevalence of technology in modernity should be understood as a central aspect of the "positivity" or "identity thinking" characteristic of the Enlightenment scientific revolution itself, characteristic, even, of the appeal to rationality throughout the Western tradition. Whereas Marx had understood science and technology as progressive forces, helping to create the material conditions for capitalism's self-overcoming, Lukács was the first to charge that science and technology also assume ideological functions in capitalism, contributing to an ideological distortion he called "reification." Horkheimer and Adorno (to a large extent developing a Nietzschean theme) radically extend this sort of critique.

What poses in modernity as a rationally enlightened attack on superstition, mythic consciousness, religion, and feudal social practices is presented by them as not only narrowing the arena of rational discourse (with great psychic costs, as in their studies of fascism), but as a form of thought incapable of, and deeply resistant to, self-critique, and as a way of linking rationality in the natural sciences and social sphere with total control and predictability, in a way that again cannot assess or reflect on the ends served by such control. Incapable of such deeper reflection, Enlightenment thus itself becomes a myth or ideology, a promotion of control or power for its own sake, to the point of pathology. (Moreover, Horkheimer and Adorno especially do not treat this connection between rationality and domination as a historical phenomenon peculiar to capitalism and the predominance of the commodity form of labor power, as Lukács does. It seems to them characteristic of all attempts at integrative rationality, as visible in the Odysseus and Sirens story as in Faust.[12] This will mean that the critical contrast with such objectification and domination will have to be a rather romantic notion of "the natural" and a proposed reconciliation with nature that is, in Kantian terms, pre-critical.[13])

Further, especially in Marcuse's account, such an ideology is far more successfully *integrative* than any previously. Thanks mainly to the culture industry, what previous critics identified as signs of strain and potential contradiction in such integrative programs—the subjective experience of alienation, lack of reconciliation with others or such a system—have been eliminated. Individuals are progressively more reconciled to social authority at a deeper, and more psychologically complex, and perhaps permanent level.

There is a famous, often mentioned problem with such accounts. The criticisms vary, depending on one's interpretation of the position; especially the extent to which one takes its proponents to be offering an indictment of the structure of modern scientific method and technology as such. An attack on the modern relation to nature (and others) as essentially a relation of domination presumes that there are alternatives models *in* the natural science tradition which could preserve the canons of objectivity, repeatability of experiment, testability of hypotheses, relatively clear confirmation relations between observation and theory, etc., but which did not embody the relation of domination. It presumes as well a model of technology not wedded to the notion of mastery of nature. It is not at all clear what a New Science or New Technology would be like.[14] (It is also not clear to what extent Marcuse is committed to such a notion.)[15] In the case of Horkheimer and especially Adorno, something like this problem produced the wholly "negative" notion of resistance so associated with their program. Simply resisting the transformation of social relations into managed, technically modeled, or bargaining relations, and of natural-aesthetic relations into manipulative, means-ends relations, seems to be touted as an end in itself. And for many such a conclusion reduces resistance to little more than a symbolic gesture.

This is of course not the final story of the approach suggested by Horkheimer, Adorno, and Marcuse. Worthy of note and much attention is a recent book by Andrew Feenberg (*Critical Theory of Technology*) which attempts to revive some of Marcuse's insights, without the unacceptable utopianism and romanticism it appeared Marcuse was committed to.[16] This involves showing the ways in which the *design*, implementation, and organization of technology are, in various ways, historically contingent, that they, at least partly, reflect the interests of "elites" who do the designing and implementing, and that some form of a democratization of the workforce can make the best, most just, social use of the now critically revealed "contingencies" in design and implementation.

None of this, he argues, requires the familiar "trade-off" between efficiency and justice sometimes said to be at stake in traditional debates. This approach historicizes the question of technology itself, such that there is no such thing as, simply, "technology," or *the* technological en-framing; there is technology designed in a certain social period for various tasks, embodying various ends, organized under certain normative assumptions.

The "critical theory of technology" promoted by Feenberg has a number of virtues. It avoids the limitations of the instrumentalist account of technology (a tool is just a tool, a hammer a hammer whether used in carpentry or to bang a tree trunk in Samoa), and the excesses of the "substantialist" approach, wherein technology embodies a world-orientation. To the former it points to the variety of contingent ways a particular technology for a particular purpose was designed, and how important political notions of administrative control and hierarchical principles of organization were inherent in such design. To the latter it makes the same point about contingency, that technologies represent complicated and often contradictory political decisions, even if it is still possible to maintain an essentially human interest in efficiency and productive power.

However, as Feenberg admits, there is no compelling reason to think that any sort of democratization of industrial organization and technology design will simply lead, thereby, to a substantially different form of production. It would certainly make such reform *possible* in ways not now possible; there are good reasons to think that various reforms enhancing autonomy, the chances of being the "subject" rather than the "object" of workplace technology, reforms enhancing interesting, diversified work and in general self-respect, are made much more difficult by the imperatives of power and control and *not* by considerations of efficiency.

But the problem here is deeper and will help introduce the larger problems facing ideology critiques. *Whoever* is in charge of the design and implementation of technology will be an agent deeply socialized in a modern ethos. And it is still not clear that such an ethos possesses the resources within it to sustain a political and ethical appeal to a reform that may result in a system just as efficient, but more humane and just. The press of the rewards now in place, and the fragmented, often unclear basis of a call to reform and to the social solidarity and sacrifice needed to implement it, may make it too risky a venture for any modern agent. Democratizing may have a fair but dispiriting result: "re-legitimation" rather than reform.

To some extent such a question is an empirical and/or a historical one, and, it seems fair enough to admit, the most a critical analysis can do is to set out the misleading or ideological character of claims about the "necessary" constraints of efficiency, or the "requirements" of technological rationality. But the problem just suggested raises a much larger issue.

Aporiai in the "Technology as Ideology" Claim

While it is true that a massive social reliance on technology can "blind" one to various social, ethical, and even potential "ontological" implications of that reliance, none of the above accounts succeed in identifying what is fundamentally "unthought" in the ever-increasing role of technology in modernization, nor what might be the implications of the changing social status of technology (our apparently declining confidence in the autonomy, or methodological purity, or even the very efficiency, of the "purposive rationality" it embodies).

Rather, one needs to understand the original social appeal of potential technological mastery as a central aspect of the *ethos of the modern revolution itself*, a revolution which is not, I have argued elsewhere,[17] essentially a bourgeois, or capitalist, or scientific, revolution. To make a very long story very short, in that context, the right sort of *doubt* to have about the nature of the social and cultural promise of technological mastery is not a doubt about a change in our fundamental ontological orientation, or about who is really and unfairly benefiting from the payoff of the promise, or who is benefiting from ossifying and reifying one stage in the historical development of technology, or whether we are becoming the objects of the forces we were the original subjects of, or whether a form of rationality has been thoughtlessly totalized, or whether technology might have been designed differently, in ways more responsive to the social needs of those who labor, etc. These are all important questions in their own right, but none is, I think, the fundamental one.

If we make a few rather vague but relatively uncontroversial assumptions (at least for the sake of the present argument), the problem will have to be posed differently. Assume simply that there is some sort of fundamental connection between the original justification of the modern revolution itself and the "mastery of nature" promise so essential to the contemporary influence of technology. To be sure, technological power can assume a different kind of importance in any number of different social situations: early industrialization, nineteenth-century American optimism; totalitarian regimes, etc. But any very general worry about the relation between technological power and such things as our understanding of nature, the nature of knowledge, or the possibility of democratic politics (the kind of things addressed by "ideology critiques") will require some attention to the uniquely modern understanding of the *necessity for* an ever-*expanding* control over the forces of nature. There could have been such a technological orientation, or a supreme political and social significance to technological power, only when such mastery seemed both necessary and possible. And understanding the conditions under which that could occur requires, I shall try to suggest, a different sort of account than is presupposed in the standard versions of ideology critique.

So, by contrast with the modern promotion of mastery,[18] the premodern emphasis on contemplation, the belief that the best regime was a matter of chance rather than human will, the insistence on an accommodation to natural *tele*, the traditional horror at the prospects of mass, collective action, are all assumptions that can be effectively countered only if the likes of Bacon and Descartes can successfully attack and undermine the bases of such claims, and then fulfill the promise of a comprehensive alternative vision, a secure, repeatable "method" capable (according to Descartes) even of challenging God's own words to Adam, and of allowing us to "enjoy without any trouble the fruits of the earth and all good things which are to be found there."[19] Making this assumption simply means that if we want to understand the relative importance of growing technological power in modernity, its significance or meaning for modern societies, we need to understand the centrality of the technological promise to the possibility of a modern *revolution*, and so to the modern rejection of antiquity, and especially to the revolutionary notion that the future can be directed and controlled by human will. (And if this is so, *challenging* the role of technology in modernity, whether in terms of the straightforward political problems noted above or in resistance to the orientation, conception of reason, alienating social relations, or false neutrality charged in ideology critiques, will require a reexamination of, assessment of, and alternative to *that* essentially modern imperative.)

For example, in traditional accounts of the function of the "legislator" or statesman, a common assumption was that one function of those who held political power was to create a common ethical sensibility among citizens. The reproducibility of a society, its ability to rally support and fend off attack, to maintain its identity over time, required an extensive political project, judiciously and wisely administered by leaders and educators with unique talents. Successful individual *self-mastery* for the citizenry was assumed to be a primary goal of political (or politico-religious) life.

One way of asking about the emergence of the technological imperative is simply to ask about the fate of such a goal when (1) under the influence of Machiavelli, among others, such a policy comes to be seen as wildly utopian, that political history itself reveals which (usually base) passions always guide human action, no matter the motives we would like or hope would be determinative; or (2) under the influence of Hobbes, the authority of the legislator's claims to "know" what virtues, or manifestations of self-mastery, are most important to promote is challenged, when a devastating epistemological attack on the foundations of traditional politico-religious authority is mounted.

Under such conditions, political life or collective action in general might come to look either impossible or possible only under radically altered expectations. The successful mastery of nature might finally make possible our being able to face the fact that a trustworthy self-mastery is simply impossible. "Unredeemed," hedonistic agents might, though, still be able to secure common goods, so long as we (1) change our expectations about those securable goods (health, security, freedom from want, the chance for a commodious life), (2) (and this is the absolutely crucial claim) are able to produce enough surplus to be able to appeal to such interests (the *only* reliable social "glue" cementing us together) and be able to "pay off," and (3) assume that our legislator need not transform or ethically educate the souls of the citizens, but can just efficiently calculate what they will do, and so "manage" well rather than "rule." The possibility of modern democracy, under the assumption that human beings are egoistic, passion-satisfying engines, would thus depend *essentially* on a qualitatively improved, "world-historical" leap in technological power. And, as often noted, if the technology of management is sophisticated enough, we might eventually not only come to expect and rely on such an egoistic, hedonistic conception of agents, but might promote and energetically encourage such activity, under the now familiar "private vices/public benefits" formula. (Or, ironically, an energetic technological optimism is required precisely because of a kind of philosophical pessimism, a great reduction in expectation about what sort of "guidance" philosophy might provide.)

Put in a different, broader way, we need to note the fact that the kind of technical power that could make possible such a new politics is itself dependent on the successful promotion of a distinct, new sort of social ethos. In the most famous and disturbing case, the productive power necessary to generate the surplus that would make modern politics possible requires a culture of consumption and acquisition, indeed a culture of ever-expanding, ever more "stimulated" consumerism. In such a context, the question of who *controls* the productive capacity, how its surplus is *distributed*, who *designs* the technology, etc., while all important issues, do not touch the fundamental problem. The links in the modernization process have to be taken in all at once: (1) the collapse of the premodern understanding of the connection between individual virtue and public life itself leads directly and unavoidably to (2) the emergence of the altered modern expectations about the narrow possibility of peaceful, coordinated activity, which in turns requires (3) new, greatly expanded technological power, itself dependent on (4) a socialization process, the "production of demand," that will itself decisively influence and constrain all modern political life. Once we understand the way in which the modern rejection of premodern politics was itself provoked by intellectual and social crises, an absolutely fundamental connection between such politics and productive or technological power, together with the "virtues" necessary to sustain it, comes into view, and appears permanently to "frame" any possible account of the significance of any new steering or distribution program for the productive forces.

In this sense, the proper question of technology would be the question of modernity itself, the question: is a distinctly modern epoch, one characterized by a radical attempt at a break with these sorts of traditional views, and by the attempt to achieve true collective self-determination, possible? Without such an enterprise, however diverse and hard to characterize, there could have been no *centrality* to technological power in modern life.

By contrast, ideology critique seems inevitably linked either to some controversial account of origins (agents' *true interests*, hidden until "reflection" exposes them), or, as we shall see shortly, to the notion of some sort of structural encroachment by one domain or "world" over another. The historical-social formation *of those interests*, especially the interest in autonomy, the control of destiny and of one's own body (originally but not exclusively embodied by property owners), and the historical sources for the growth of such "domains" (especially the historical reasons for the collapse of traditional, teleological worldviews, the emergence of "instrumentalist" models as the only publicly defensible notion of rationality), are, I want to claim, different and more important issues.

Or: the right metaphor for understanding the extraordinary and potentially distorting appeal of technological power in modernity is not a hunt for hidden origins, or a delineation of geographical boundaries, but attention to the *context*, the historical moment when mastery in general would have seemed, with some historical urgency, an unavoidable *desideratum*. If, I want to claim, we can understand the rising importance of technological mastery in such a broader context, we will be less inclined to see that rise as some sort of Faustian bargain, prompted by hubris, narrow class interest, confusion about different domains of rationality, or as a lust for power. Ideology critique tends toward such explanations of *why* what is now claimed to be hidden or unnoticed got to be hidden or unnoticed, and, while I cannot demonstrate the claim here, I want at least to suggest that such interpretations are implausible.

Still more simply put, the modern claim that the highest, publicly defensible good is a technically efficient mastery of nature, with all its implications for social organization, ethical relations, and public life, may be, in both historical and general terms, *rational.* Modern agents might not be confused by the implications of the "philosophy of the subject," their preferences might not have been wholly formed in a situation distorted by the influence of money and power, they may not have confused the ends of work and those of interaction, they may not be falsely universalizing or naturalizing a particular historical epoch. At least we should not beg any questions in making such claims, and that will require the broader assessment I am suggesting, an assessment of the resources *within modernity* for understanding the possible narrowness of such a conception of a rational end, and the exclusion of others.

Such a broader view will also permit a more adequate understanding of the implications of the shifting social status of science and technology, away from a privileged center, under the weight of various historicist, sociological, naturalist, and other critiques, something not well understood in the rather jejune contemporary fascination with a possible postmodernism.

All of which seems to raise the stakes for understanding the role of science and technology in modern life to an unsatisfiable level (the old Hegelian "you've got to understand everything to understand anything" problem). In order to motivate this way of looking at things, I want at least to defend the claim about the limitations of the ideological approach and hope the alternative I have in mind will begin to emerge. My example will be Habermas's well-known argument.

Habermas on Technology as Ideology

According to Habermas's famous account, the great problem with technological modernization has little to do with technology itself, but with the way in which the influence, scope, and success of technology in modern life have tended to authorize only "purposive" or instrumental and strategic notions of *rationality*, and to de-legitimate (as unresolvable and subjective) genuinely practical or political questions.[20] The imperatives of "work," rational if efficient, efficient if productive, and "interaction," rational if the norms of successful, genuine communication are realized, have become confused and the specific form of rationality inherent in communicative action has been overwhelmed by the demands of technical efficiency. The "lifeworld" has been "colonized," "steered" or encroached on by the demands of a self-regulating system.

This critique, according to which it is the totalization of the instrumental notion of rationality, its absorbing the categorically distinct forms of communicative rationality, which renders science and technology ideological, appears a sensible response to the historical limitations of Marx's original account, and appropriately cognizant of the undeniable benefits of technological power over nature and our own fate.

The account is largely motivated by a critical appropriation of Max Weber's original account of modernization as the progressive rationalization and so demystification of various spheres or subsystems of modern life. On this account, control "from above," legitimated by appeal to cultural, religious, and mythic world views, constrains the development of purposive rationality in various spheres of life. The efficient satisfaction of basic needs, creation of surplus and so leisure and luxury, etc., are all impeded by such cultural (and basically irrational) constraints. This begins to change with capitalism and the creation of a purposive-rational system that demands its own continual expansion. Under the growing pressure of such expansion, traditional worldviews are transformed into private beliefs, incapable of functioning as universal or culturally stable forms of social authority, and the legitimacy of the capitalist claim to productive efficiency and universal satisfaction of interests wins out in a kind of competition for social power. The West is modernized.

This success, however—a real advance when measured against "the systematically distorted communication" and the "fateful causality of dissociated symbols and suppressed motives" of premodern forms of communicative interaction—becomes itself repressive and ideological when it prevents any *re-establishment* of a genuinely inter-active life among modern subjects, which is what Habermas claims happens, and why he claims science and technology (in their presumption to be definitive of rationality as such) are ideological.

Habermas's approach generates the following problems, all of which return us to the general issues about modernization raised above. First, Habermas does not treat the extraordinary acceleration of technological progress in modernity or the modern reliance on technology as unique historical phenomena. Following Gehlen, he interprets this as an extension of the basic structure of all human purposive-rational action. We are simply getting much better at "aiding" *homo faber* in what he has always been interested in doing: moving about better, seeing and hearing better, producing and regulating energy, governing our actions more efficiently, etc. Work is a permanent, constitutive "human interest."[21]

But this claim leaves unanswered the question: *why* (to use Habermas's language) work or the imperatives of purposive-rational activity became so much *more extraordinarily important* in the modern age, why the ideal of mastery began to occupy a qualitatively different position on the social agenda.[22] Put in terms of Descartes's rhetoric: we need to know why (rather suddenly in historical terms) we should have turned so *much* of our energy to the "*mastery* of nature." That image suggests not merely an *extension* of our human interest in successful purposive action. Even viewed within the domain of work, of being able to get done what we want to get done (ignoring for the moment whether this involves a wholesale new relation to nature, or world-orientation, or understanding of Being), this image suggests a kind of urgency, a situation of insecurity requiring a military assault against an *enemy*, all not captured in Habermas's account.[23]

Second, Habermas supplements Weber's account of rationalization with his own picture of the competing requirements of work and communicative or inter-active activity. He suggests a theory of modernity in which the premodern standards of communicative interaction, since they were prejudiced, distorted products of repression, etc., "gave way" under the press of the successful expansion of productive capacity and the purposive-rational standards of rationality that go with it. It is unclear whether this account is limited to certain social and economic aspects of modernization, or is meant to identify the basic origin of modernization. The latter is suggested by the frequent references to Weber, and to the larger intellectual, philosophical, and religious issues supposedly called into question by the "expansion" of purposive standards of rationality into numerous subsystems.

Such an account, in the first place, downplays the philosophical crisis brewing in the tradition since nominalism. It would be hard to understand the *promotion* of utility, *why* we should become interested in becoming "masters and possessors of nature," should have displaced the notion of knowledge as contemplation, or explored an account of knowledge (the "new way of ideas") subject to methodological rigor and control, unless such a philosophical tradition is taken into account.[24] For example, the importance of developments in technology could not possibly have risen to such a high spot on the social agenda without the essentially modern view that the source of most if not all human misery was *scarcity* (a new and quite controversial claim) and that scarcity was a solvable technical problem, nor without a new notion of "knowing as making," inspired by developments in mathematics.[25] The simple emergence of a new, more optimistic view of how much of the basic scarcity problem could be technically resolved would not be a deeply significant discovery had the classical view still held sway: that the central political problem is an unimprovable finitude, a basic, permanent distinction between the few and the many, and the unpredictable, wholly contingent congruence of wisdom and political power.

And the metaphor of one view pressing on or pushing aside another is not very persuasive. We need an account and an assessment of such a new view of the basic problem. Why did it arise when it did? Was it a rational thing then to believe? For whom? Under what conditions? The general picture that Habermas paints, with one version of rationality in a kind of boundary war with another, does not adequately account for the historical context in which they would have first been seen as competitors, and in which their competition would have been seen as significant.[26]

The essential point is this: without a sufficient understanding of that larger context, we shall be unable to understand the *consequences* of any sort of acknowledgment of the "limitations" of instrumental rationality. It wouldn't matter if such limitations are claimed because of a growing tendency to historicize scientific and technological procedure, or to question the neutral, universalist pretensions of such procedures (by attention to sociological, psychological, or pragmatic origins), or as a result of a critical attack on the ideological totalization of purposive rationality. We shall not be in a position to understand what such limitations amount to without quite a broad view of the historical landscape.

If, for example, what Habermas calls the displacement of practical by technical questions occurred because *any* version of practical politics must appeal to essentially premodern and no longer defensible notions of ends, teleology, nature, etc., or can be resolved only by a kind of strategic bargaining among agents with incommensurable goals, then there are no significant practical consequences from such claims about limitations. There just *would be no possible agenda* for the practical realm, and a growing lack of confidence in the standard or traditional claims of science and technology to provide unprejudiced or neutral means to satisfy any sort of ends would only result in a *greater* skepticism and social fragmentation, *not* emancipation.

Now Habermas claims in his own voice that there is such an alternative agenda, one based on freeing communication from arbitrarily imposed limits and distortions caused by the interests of money and power, and so promoting an "ideal speech situation." This claim has given rise to objections that Habermas's distinction between purposive and communicative rationality is often arbitrarily drawn. On such a view, Habermas has a curiously positivistic understanding of "science and technology," as if they really are, if restricted to the proper sphere, as squeaky clean methodologically as traditionally maintained.[27] Many recent discussions in philosophy of science, from historical and sociological studies of scientific practice to issues raised by a new generation of scientific realists, have created a number of doubts about the possibility of such boundaries, and at the very least raise as a possibility that there is no methodological way to isolate the purposive-rational dimension of science and technology and preserve it safely in its own domain. Admitting that the institution of science, its organization, hierarchy, criteria of success, or criteria of good or central, as opposed to bad or marginal, science, etc., and the design and implementation of technology are everywhere *already*, themselves, "symbolically mediated," that social values interpenetrate at every level, helping to define purposive effectiveness, need not mean we are committed to some contrary or new science. We may just have successfully pointed out the ever "embodied" nature of thought and the severe problems Habermas faces trying to keep things in their proper boxes.

Habermas faces a similar problem if it can be shown that the significance of the modern emphasis on the form of rationality embodied in science and technology cannot be explained as a result of some sort of contingent (and reversible) displacement of inter-active norms by purposive-rational norms. If there is some more comprehensive historical context within which both the abandonment of traditional value systems and the allegiance to an instrumental notion of rationality could be explained and motivated, then promoting open allegiance to rules of ideal communicative equality, within such a context or whole, might and likely would simply institute a formally fair way of *re-legitimating* the substantial anomie, fragmentation and dissatisfactions of modernity, forcing us, after our freely arrived at and communicatively fair interaction, back to the narrow confines of strategic and instrumental rationality as the best concrete, realizable hope we've got for coordination.

Habermas of course disagrees, thinking, first, that there would be some sort of pragmatic contradiction in a situation where agents, even while conforming to ideal speech conditions, sanctioned instrumental social relations and power relationships justified only by instrumental efficiency, and second, that a very great deal in modern social life would change if such an ideal speech situation were achieved. And his account is more sensitive to issues and more nuanced than I have been able to present it in this summary. In many ways it remains a powerfully critical approach to many aspects of modernization. But I do not think the basic strategy—the "separate into relevant spheres" approach—goes deep enough into the modern origins of such original separations, and so leaves too unclear the implications of the Habermasian delimitations.

The most general and now quite familiar way to state these issues would be to say that Habermas and the whole ideology-critique tradition still remain bound by the Kantian assumptions which made it possible, still wedded to the hope that a formal account of communicative practice, or the conditions of interest formation, or of reflexivity in general, will provide us with the critical tools necessary to understand modernity. My claim has been, not that such a direction is misguided,[28] but that these issues need to be raised within a broader philosophic framework, one more sensitive to the substantive, historical, and practical issues at stake in the modern revolution itself, and so more responsive to the claim that modernity, and its technological implications, is, finally and decisively, "legitimate"; at the very least, in historical terms, "sufficiently rational."[29]

Notes and References

1. Cf. Hannah Arendt, *The Human Condition* (Chicago: University of Chicago Press, 1958); Jürgen Habermas, *Strukturwandel der Öffentlichkeit* (Berlin: Leuchterhand, 1962).

2. Cf. *inter alia*, the essays in Michel Foucault, *Power/Knowledge: Selected Interviews and Other Writings*, ed. C. Gordon (New York: Pantheon, 1980).

3. Cf. the historical account in Norman Stockman, *Antipositivist Theories of the Sciences* (Dordrecht: Reidel, 1983), Section 3.2, "Critical Theory's Critique of Positivism: The Kantian Background," pp. 43–51. Stockman's account should be supplemented by additional attention to the role of Lukács. See Andrew Feenberg, *Lukács, Marx and the Sources of Critical Theory* (Oxford: Oxford University Press, 1981).

4. See the three types of "pejorative" ideology critiques identified by Raymond Geuss in *The Idea of a Critical Theory: Habermas and the Frankfurt School* (Cambridge: Cambridge University Press, 1981), p. 13.

5. This sort of theme appears in many of Heidegger's later writings, but I shall treat as typical such essays as "The Age of the World Picture," "The Word of Nietzsche: God Is Dead," and "The Turning," in *The Question Concerning Technology and Other Essays* (New York: Harper and Row,

1977), and the concluding lectures in the Nietzsche series, *Nietzsche, Vol.IV. Nihilism*, trans. D. Krell (San Francisco: Harper and Row, 1982).

6. Jacques Ellul, *The Technological Society*, trans. J. Neugroschel (New York: Continuum, 1980).

7. Robert B. Pippin, *Modernism as a Philosophical Problem: On the Dissatisfactions of European High Culture* (Oxford: Basil Blackwell, 1991), pp. 117–47.

8. See the references above in n.2, and the useful discussion in David Hoy, "Power, Repression, Progress: Foucault, Lukes, and the Frankfurt School," in *Foucault: A Critical Reader* (Oxford: Basil Blackwell, 1986), pp. 123–48.

9. For representative passages, see Karl Marx, *Capital*, ed. F. Engels, trans. Samuel Moore and Edward Aveling (New York: International Publishers, 1977), pp. 312–507; volume I, Part IV, "Production of Relative Surplus Value," especially chapter 15, "Machinery and Modern Industry." For the contrast between production under capitalism and after, see *Grundrisse*, trans. Martin Nicolaus (New York: Vintage, 1973), p. 488.

10. M. Horkheimer, and T. Adorno, *The Dialectic of Enlightenment* (New York: Seabury, 1972).

11. Herbert Marcuse, *One-Dimensional Man* (Boston: Beacon Press, 1964).

12. Cf. Max Horkheimer, *Eclipse of Reason* (New York: Seabury, 1974), p. 176.

13. This is of course not true of the position developed by Adorno in *Negative Dialectics*, but that is a longer story. See my discussion in *Modernism as a Philosphical Problem*, op.cit., pp. 151–56, and Habermas's statement of his differences with the Horkheimer-Adorno approach in *The Theory of Communicative Action*, trans. T. McCarthy (Boston: Beacon, 1984) vol. I, IV.2, "The Critique of Instrumental Reason," pp. 366–99; and in *The Philosophical Discourse of Modernity*, trans. F. Lawrence (Cambridge, MA: MIT Press, 1987), chapter 5, pp. 106–30.

14. Cf. Stockman, op.cit., pp. 57–64, and 240–46.

15. Marcuse, *One-Dimensional Man*, op.cit., p. 154.

16. Andrew Feenberg, *Critical Theory of Technology* (Oxford: Oxford University Press, 1991).

17. Robert B. Pippin, *Modernism as a Philosophical Problem*, op.cit.

18. I shall be assuming that the modern enterprise does not merely seize the opportunity to *extend* a "natural," species-characteristic interest in the control of nature through labor and tools, but that the early moderns began to reformulate the range of natural events that could be mastered, what could count as such mastery (given the new influence of mathematics, and a new attention to the problem of certainty), what such mastery was *for*, and the relation between such a goal and other desirable social ends. Given such a claim, understanding why such reformulations occurred cannot be answered by appeal to a mere extension of such a species characteristic interest. I am thus disagreeing with, e.g., Habermas's account. See the discussion below.

19. René Descartes, *Discourse on Method*, in *The Philosophical Works of Descartes*, trans. Elizabeth Haldane and G. R. T. Ross (Cambridge: Cambridge University Press, 1969), vol. I, pp. 119–20. Cf. *Genesis*, chapter 3, verse 17.

20. Especially in J. Habermas, "Technology and Science as Ideology," in *Toward a Rational Society*, trans. J. Shapiro (Boston: Beacon Press, 1970), and in Vol. II of *The Theory of Communicative Action*, trans. T. McCarthy (Boston: Beacon Press, 1987).

21. Habermas, "Technology and Science as Ideology," op.cit., p. 87.

22. Cf. for example his account of how "pressure" from the development of productive forces brings about, as if by hydraulic force, the end of traditional societies, "Technology and Science as Ideology," p. 96.

23. Contrast the different account of the significance and unique characteristics of labor in the modern world in Arendt, *The Human Condition* op.cit., chapter 3.

24. Among the many studies of this complicated intellectual development, see Hans Blumenberg, *The Legitimacy of the Modern Age*, trans. Robert Wallace (Cambridge, MA: MIT Press, 1983), and Amos Funkenstein, *Theology and the Scientific Imagination from the Middle Ages to the Seventeenth Century* (Princeton: Princeton University Press, 1986). See also my "Blumenberg and the

Modernity Problem," *Review of Metaphysics* 40 (1987): 535–57, and chapter 2 of *Modernism as a Philosophical Problem*, op.cit.

25. Cf. the account in David Lachterman, *The Ethics of Geometry: A Genealogy of Modernity* (New York: Routledge, 1989).

26. See the criticisms of Habermas by Axel Honneth in *Kritik der Macht: Reflexionstufen einer kritischen Gesellschaftstheorie* (Frankfurt: Suhrkamp, 1985), and my "Hegel, Modernity and Habermas," *Monist* 74 (July 1991): 329–57.

27. Cf. Stockman's account, op.cit., pp. 109–12.

28. Cf. the development of this theme in my "Marcuse on Hegel and Historicity," in *Marcuse: Critical Theory and the Promise of Utopia*, ed. R. Pippin, A. Feenberg, C. Webel (London: Macmillan, 1988).

29. I am thinking here of Hans Blumenberg's strategy and terms. See his *The Legitimacy of the Modern Age*, op.cit.

PART II

Technology and the Moral Order

PIPPIN'S CHAPTER SUGGESTS that the crisis of technology is really a crisis in the Western moral tradition. The two chapters of this section fault that tradition for failing to supply us with the conceptual means to understand and control technology effectively. Only a significant revision of traditional views can restore practical thinking to a central place in our culture.

In "Citizen Virtues in a Technological Order," Langdon Winner notes that the Western ethical tradition has consistently separated technology and the public sphere. Ancient philosophy exalted public life while looking down on the practical arts and technology. Modern philosophy learned to appreciate the value of technique, but carefully assigned it to the private sphere, once again excluding it from politics. No wonder, Winner argues, that citizenship still does not encompass technology despite the overwhelming role of technical choices in modern societies. A new concept of citizenship is needed, exercised in new public spaces for debating technical change. Scandinavian experiments in participatory technological design offer an innovative approach to restoring public life.

Albert Borgmann's "The Moral Significance of the Material Culture" makes a similar point. Philosophy has never achieved a proper balance between theory, practice, and material life. Modern philosophy emphasizes theory at the expense of practice, and while postmodern thought begins to recognize the importance of practice, it continues to underestimate the significance of the material domain. As a result, philosophy is completely disarmed by the new moral choices we face in a technological society. It is time to begin evaluating the qualitative differences between different types of goods and the different styles of material life which they support, much as we evaluate ethical alternatives. Research on the place of objects in the home suggests an approach to what Borgmann calls "real ethics." As an example, he compares the very different way of life implied in the intense involvement of the user with musical instruments and the passive experience of listening to a stereo.

4

Citizen Virtues in a Technological Order

Langdon Winner

A s it ponders important social choices that involve the application of new tech-nology, contemporary moral philosophy works within a vacuum. The vacuum is created, in large part, by an absence of widely shared understandings, reasons, and perspectives that might guide societies as they confront the powers offered by new machines, techniques, and large-scale technological systems. Which computer applications are desirable and which ought to be avoided? How can one weigh the risks of introducing a new chemical into the environment as compared to benefits of its use? Should there be limits placed upon the ability of biotechnology to alter the genetic structure of plant and animal life? As we ponder issues of this kind, it is not always clear which principles, policies, or forms of moral reasoning are suited to the choices at hand.

The vacuum is a social as well as intellectual one. Often there are no persons or organizations with clear authority to make the decisions that matter. In fact, there may be no clearly defined social channels in which important moral issues can be ad-dressed at all. Typically, what happens in such cases is that, as time passes, a mixture of corporate plans, market choices, interest group activities, lawsuits, and government legislation takes shape to produce jerrybuilt policies. But given the number of points at which technologies generate significant social stress and conflict, this familiar pat-tern is increasingly unsatisfactory.

Philosophers sometimes rush in to fill the void, offering advice that matches their training and competence. They examine cases in which some feature of a present or emerging technology raises questions about right and wrong in individual choices and social policies. They take note of properties of the new technology that have important consequences for social life, properties that raise interesting philosophical issues; for example, issues about the rights and responsibilities of those who develop or use the technology in question. From there they can develop a variety of theories, principles, and arguments that may help people decide what to do.

Proceeding in this way, philosophers may find themselves involved in an exercise that is essentially technocratic. The complicated business of research, development, and application in modern life includes a moment where the "value issues" need to be studied and where the contributions of knowledgeable, degree-carrying experts can be enlisted. In the United States, for example, the National Science Foundation has for many years included a program on "ethical and value studies" that supports university scholars who do research of this kind. The underlying assumption seems to be that this is an important area that the nation needs to cultivate. The sponsors may hope that officially designated "values experts" can eventually provide "solutions" to the kinds of "problems" whose features are ethical rather than solely technical. This can serve as a final tune-up for working technological models about to be rolled out the showroom door. "Everything else looks good. What are the results from the ethics lab?"

Philosophers sometimes find it tempting to play along with these expectations, gratifying to find that anyone cares about what they think, exhilarating to notice that their ideas might actually have some effect. But is it wise to don the mantle of values expert? Although philosophers may be well equipped to help fill the intellectual emptiness caused by the lack of moral understandings, ethical reasoning, and community guidelines, there remains the social and political vacuum that so often surrounds discussions about the moral dimensions of technological choice. After one has addressed the range of social theories, empirical analyses, philosophical arguments, and ethical principles about the possibilities of Technology *X*, there remains the embarrassing question: Who in the world are we talking to? Where is the community in which our wisdom will be welcome?

Consider the following passages from two prominent writers addressing urgent ethical questions for our time. The first is from a well-known biologist reflecting about the ethical dimensions of developments in his own field.

> Given the nature of our society, which embraces and applies any new technology, it appears that there is no means, short of unwanted catastrophe, to prevent the development of [human] genetic engineering. It will proceed. But this time, perhaps we can seek to anticipate and guide its consequences.[1]

The second passage was written by a professional philosopher, exploring avenues for the new field of computer ethics.

> We are open to invisible abuse or invisible programming of inappropriate values or invisible miscalculation. The challenge for computer ethics is to formulate policies which will help us deal with this dilemma. We must decide when to trust computers and when not to trust them.[2]

Both of these passages are notable for the way they employ the term *we* in contexts where moral issues about technology are open for discussion. But who is the "we" to whom the writers refer? Both writers seem to mean something like "people in general" or "society as a whole." Or perhaps they mean something like "those who work in a particular field of technical development and have privileged access to the decisions that matter."

I raise this point not to call attention to the way writers, including this one, loosely deploy first-person plural pronouns. What matters here is that this lovely "we" suggests the presence of a moral community that may not, in fact, exist at all, at least not in any coherent, self-conscious form. If "we" scholars find ourselves talking about a collectivity of others who are not in fact engaged in decisions, then it is time for "us" to look around and find out where "they" have gone. That is the important first task for the contemporary ethics of technology. It is time to ask: what is the identity and character of the moral communities that will make the crucial, world-altering judgments and take appropriate action as a result?

This question is, in my view, one about politics and political philosophy rather than a question for ethics considered solely as a matter of right and wrong in individual conduct. For the central issues here concern how the members of society manage their common affairs and seek the common good. Because technological things so often become central features in widely shared arrangements and conditions of life in contemporary society, there is an urgent need to think about them in a political light. Rather than continue the technocratic pattern in which philosophers advise a narrowly defined set of decision makers about ethical subtleties, today's thinkers would do better to reexamine the role of the public in matters of this kind. How can and should democratic citizenry participate in decision making about technology?

Unfortunately, the Western tradition of moral and political philosophy has little to recommend on this score, almost nothing to say about the ways in which persons in their roles as citizens might be involved in making choices about the development, deployment, and use of new technology. Most thinkers in our tradition have placed technology and politics in separate categories, defining citizen roles as completely isolated from the realities of technical practice and technical change. There have been two distinctive paths to this conclusion, one characteristic of thinkers in antiquity, another strongly advanced in modern times. But whether we are pondering ancient *techne* or today's megatechnics, any attempt to discuss technology as a topic in political and moral philosophy needs to pause long enough to appreciate how this crucial separation occurred and how it impairs our sense of possibilities.

Technology and Citizen: The Ancient View

At the beginning of Western moral and political philosophy, speculation about *techne*, the realm of the practical arts, plays a prominent but largely negative role. As Socrates, Plato, and Aristotle seek to define the nature of knowledge, the good, political society, justice, rulers and citizens, and the form of the best state, they frequently draw comparisons to *techne*, the realm of the arts and crafts, viewing it with a mixture of awe and suspicion. Foremost among their concerns is the belief that technical affairs constitute an inferior realm of objects, knowledge, and practice, one that threatens to infect all who aspire to higher things.[3] Plato goes even further, specifying why the realm of *techne* is both inferior and potentially dangerous. True knowledge, he argues, is not that of worldly, mutable, material things, but knowledge of the realm of unchanging ideas, *eidos*.[4]

Arguing a position that was to become commonplace in antiquity and throughout much of the Middle Ages, Plato also criticizes the practical arts for their tendency to produce innovations, a source of harmful, potentially boundless change in human affairs. Political philosophy seeks to establish good order and to maintain it against the world's tendency toward chaos and decay. "Change, we shall find, is much the most dangerous thing in everything except what is bad—in all the seasons, in bodily habits, and in the characters of souls."[5] In the first century B.C. Lucretius echoes these sentiments, lamenting the destructive role of new techniques in warfare. "Tragic discord gave birth to one invention after another for the intimidation of the nations' fighting men and added daily increments to the horrors of war."[6]

Of all classical arguments calling for the separation of technology from political affairs, the most significant is Aristotle's. For unlike Plato, Aristotle explores the possibilities of a broadly based citizenship in political societies of many different kinds, perhaps even ones that resemble our own. As he defines the roles and virtues of a citizen, however, the crucial differences between technical and political life stand out.

Aristotle's view that "man is by nature a political animal" means that humans are creatures naturally suited to live in a *polis* or city-state.[7] Drawing upon studies of some one hundred and fifty city-states of his time, the *Politics* argues that the *polis* is the highest form of human organization, one that completes the development of other forms of association, the household and the village. Political life is a gathering of freemen and equals. Each person is free in the sense that there is no master to dictate one's activities. Each one is equal as well, equal in legal standing, access to public office, and right to speak in political matters. Political life concerns matters that all citizens have in common. In the public sphere one's attention moves beyond personal or family interests to seek the good of the whole community. "One citizen differs from another, but the salvation of the community is the common business of them all."[8] Citizenship, active participation in public life, fulfills man's highest potential. The *bios politicos* realizes a greater good than more primitive forms of human existence ever attain.

Having defined politics in this manner, Aristotle goes on to explore the specific roles and virtues of the citizen. He notes the traditional distinction between the rulers and the ruled and concludes that the citizen must be different from both. Citizenship in his view must include both roles within each person.[9]

> The excellence of the two is not the same, but the good citizen ought to be capable of both; he should know how to govern like a freeman, and how to obey like a freeman—these are the excellences of a citizen. And although the temperance and justice of a ruler are distinct from those of a subject, the excellence of a good man will include both. . . .

Looking at a range of existing constitutions, Aristotle concludes that a good constitution will allow the rotation of citizens in office so the "excellences" or "virtues" he recommends will become common in actual practice.

In the same passages that offer his definition of citizenship, Aristotle takes care to specify which persons are not capable of holding this role. He points to the menial duties and craft work that were handled by slaves and foreign workers in Greek city-states of the time. Physical toil and use of the practical arts bind one to the realm of material necessity, a condition incompatible with the unencumbered freedom needed for citizenship. While slaves and craftsmen are necessary for the existence of the state and while some city-states recognize them as citizens, a good society will not extend citizenship in this way, "for no man can practice excellence who is living the life of a mechanic or labourer."[10]

Aristotle goes even further, arguing that citizens should avoid learning the practical arts because that would be degrading. "Certainly the good man and the statesman and the good citizen ought not to learn the crafts of inferiors except for their own occasional use; if they habitually practice them, there will cease to be a distinction between master and slave."[11] Thus, the making of useful things and the activities of public life must forever remain separate.

While the ideas of Socrates, Plato, and Aristotle did not by themselves define the understanding of the Greeks and Romans on such matters, entirely similar notions about technology and economics were common in antiquity. The sphere of technical affairs was closely associated with slavery and menial labor and was, therefore, something that persons of the ruling classes sought to avoid. In fact wealthy Romans normally left the day-to-day handling of private economic affairs to their slaves, the origins of what we today call "management."[12] While Romans sought material wealth, it was usually gained through landed property and commercial trade, economic sources that did not require recurring technical change. Indeed, technological innovation was widely regarded with suspicion. Suetonius tells of a time when a creative soul came to the emperor Vespasian with a device for carrying heavy columns into Rome at a low cost. Although Vespasian rewarded the man for his invention, he refused to use it, exclaiming, "How will it be possible for me to feed the populace?"[13] As the historian M. I. Finley concludes in *The Ancient Economy*, "Economic growth, technical progress, increasing efficiency are not 'natural' virtues; they have not always been possibilities or even desiderata, at least not for those who controlled the means by which to try to achieve them."[14]

Technology and Citizen: The Modern View

With the renewal of political theory in the sixteenth century and since, the prospects for social and political life are gradually redefined. Concepts of power, authority, order, liberty, equality, and the state are deployed in ways that we now consider distinctly modern. The attempts of Machiavelli, More, Hobbes, Locke, Montesquieu, Bentham, and Marx to create a new understanding of politics corresponded to path-breaking work in natural sciences that produced new ways of thinking about the physical world. Strongly associated with these intellectual movements is a thoroughgoing reevaluation of the sphere of technical practice and its economic settings, a reevaluation in which the pessimism of ancient and medieval views eventually yields to an unbridled optimism. In this ferment of ideas, the traditional view of the relationship between politics and technology was overthrown and a new one imagined.

A leader in promoting respect for technical activity was Francis Bacon. In *The New Organon* Bacon surveys the state of knowledge in his time, criticizing the hold of the ancient philosophers over the minds of moderns. He argues that the supposed wisdom of the Greeks is suspect precisely because it lacks any practical, material value: "it can talk, but it cannot generate, for it is fruitful of controversies but barren of works."[15] As an alternative Bacon sets forth a new program of knowledge and practice, one based upon careful study of particular phenomena, adherence to method, inductive logic, controlled experiment, naturalistic explanation, and a specialized division of labor among scientists. The ultimate purpose of such activity, he makes clear, ought to be the conquest of nature and expansion of human powers. Natural philosophy must go beyond the quest for knowledge as an end in itself and seek fulfillment in the practical arts.

As a former politician who had fallen from power in disgrace, Bacon enthusiastically praises the superiority of the new scientific and technical pursuits in contrast to affairs of state. Comparing the contributions of history's political heroes to those who have made wonderful discoveries and inventions, Bacon concludes that the highest honors go to scientific and technical innovators, "For the benefits of discoveries may extend to the whole race of man, civil benefits only to particular places; the latter last not beyond a few ages, the former through all time."[16]

Although Bacon's expectations about the directions the arts and sciences ought to pursue were not always prescient, his promotional views won numerous followers in later generations. Explicitly taking his advice, many French philosophers of the eighteenth century took great care to stress not only the practical value of technical pursuits but their intellectual strengths as well. In his *Preliminary Discourse to the Encyclopedia of Diderot,* Jean Le Rond D'Alembert notes the widespread contempt that surrounds the mechanical arts, an outlook that even the artisans themselves seem to share. He argues that, in fact, "it is perhaps in the artisan that one must seek the most admirable evidences of the sagacity, the patience, and the resources of the mind."[17]

Closely linked to a more favorable view of the practical arts and technical innovation is a change in attitude toward commerce and material self-interest. During the Middle Ages, avarice was often identified as both a sin and a source of civil unrest. While medieval societies were often quite open in their quest for wealth, the dominant view among church, political, and intellectual elites was that such motives should be carefully contained. A significant development in modern social and political thought was to annul this distrust and to recast ideas about wealth and commerce in an entirely favorable light. The pursuit of economic gain, some philosophers began to argue, is actually a force for moderation, helping to nurture more rational, peace-loving attitudes among both rulers and subjects. Persons with an economic stake in such trade and manufacturing were now thought to be healthy contributors to stability and justice in political society.[18] As Baron de Montesquieu argues in *The Spirit of the Laws*, "the spirit of commerce is naturally attended with that of frugality, economy, moderation, labor, prudence, tranquillity, order, and rule. So as long as this spirit subsists, the riches it produces have no bad effect."[19] Commerce, he argues, has another beneficial effect, binding nations together in a pattern of mutual need that discourages conflict.

Ideas of this sort, increasingly common in seventeenth- and eighteenth-century political theories, helped justify the modern optimism about economic self-interest and faith in the beneficence of economic growth which lie at the foundation of modern liberal thought. In the new understanding, wealth is good not only for its material benefits but also because its pursuit produces better rulers and better citizens.

The idea that self-interested economic activity is fundamental to politics is strongly expressed in the writings of John Locke. In *The Second Treatise of Government*, Locke's conception of man is that of an acquisitive creature who subdues nature and makes it his property. Men leave the "state of nature" when they come to realize that their possessions are insecure. They form a society and, as a second step, submit to the rule of a government which recognizes their rights, particularly the right of property. From this point of view, the function of political society and government is that of defending the holdings of what are in essence private individuals. If it turns out that government is not useful in achieving these purposes, it can be rightfully overturned in revolution.

At the center of Locke's theory of political society and of modern liberal theory in general is a conception of human life that C. B. MacPherson has called "possessive individualism."[20] In this vision, acquisitiveness emerges as a positive, civilizing force. For as people pursue material gain, they become more rational, industrious, peaceful, and law-abiding. Hence the purely private virtues appropriate to a market society and capitalism are the virtues that build a stable political order. Of the activities that help produce a good society, none are superior to technical pursuits. As David Hume explains in his essay "Of Refinement in the Arts," "In times when industry and the arts flourish, men are kept in perpetual occupation, and enjoy, as their reward, the occupation itself, as well as those pleasures which are the fruit of their labour. The mind acquires new vigour; enlarges its powers and faculties. . . . "[21] For that reason Hume advises rulers to encourage the development of manufacturing even in preference to agriculture. Dynamic new enterprises are more civilizing than the bucolic traditions of farming.

An important feature of this persuasion in contrast to classical notions is that poli-

tics is assigned a relatively low position in the broader scheme of human affairs. For
Locke, government is an instrument with no intrinsic value. Its role is to protect the
rights of "life, liberty, and property" by serving as an umpire when disputes arise. Attending to governmental matters is certainly not a sphere in which a person can realize
one's highest potential. Locke finds no higher meaning in the realm of citizen action.
One enters the public realm merely to express one's private interests. In contrast to
Aristotle's view, Lockean liberalism recognizes neither goods nor virtues that stem from
one's being as a public person.

In *The Wealth of Nations* Adam Smith develops the belief in the primacy of private
affairs to its logical conclusion, viewing all public interference with scorn. Government measures, he argues, have "retarded the natural progress of England towards
wealth and improvement. . . ."[22] Government is the source of extravagance, miscon-
duct, and countless ill-conceived projects while the "uniform, constant and uninter-
rupted effort of every man to better his condition"[23] he identifies as the wellspring of
most private and public good.

> It is the highest impertinence and presumption, therefore, in kings and minis-
> ters to pretend to watch over the economy of private people, and to restrain
> their expense, either by sumptuary laws, or by prohibiting the importation of
> foreign luxuries. They are themselves always, and without any exception, the
> greatest spendthrifts in the society. Let them look well after their own expense,
> and they may safely trust private people with theirs.[24]

Ideas of this kind underlie basic institutions of politics and economics in modern
liberal democracies, posing strong barriers to attempts to think about the public dimensions of technological choice. Technological change, defined as "progress," is
seen as an ineluctable process in modern history, one that develops as the result of the
activities of men and women seeking private good, activities which include the development of inventions and innovations that benefit all of society. To encourage progress
is to encourage private inventors and entrepreneurs to work unimpeded by state interference. As later theorists in the liberal tradition modify this understanding, they notice
"market externalities" that cause stress in the social system or environment. This does
not alter the fundamental attitude toward economic and technical choices. The burden of proof rests on those who would interfere with beneficent workings of the market
and processes of technological development.

If one compares liberal ideology about politics and technology with its classical
precursors, an interesting irony emerges. In modern thought the ancient pessimism
about *techne* is eventually replaced by all-out enthusiasm for technological advance.
At the same time basic conceptions of politics and political membership are reformulated in ways that help create new contexts for the exercise of power and authority.
Despite the radical thrust of these intellectual developments, however, the classical
separation between the political and the technical spheres is strongly preserved, but

for entirely new reasons. Technology is still isolated from public life in both principle and practice. Citizens are strongly encouraged to become involved in improving modern material culture, but only in the market or other highly privatized settings. There is no moral community or public space in which technological issues are topics for deliberation, debate, and shared action.

Technology and the Quality of Contemporary Citizenship

The hollowness of modern citizenship, the paucity of citizen roles and lack of opportunities for direct participation in politics, is now a general condition, not limited to technology policy-making alone. Many writers have lamented structures of representative democracy that effectively exclude ordinary people from significant involvement in public affairs. Thus, Hannah Arendt notes with approval Thomas Jefferson's proposals that American government include "elementary republics" that might have brought small-scale political assemblies into the realm of everyday life. "What he perceived to be the mortal danger to the republic was that the Constitution had given all power to the citizens, without giving them the opportunity of *being* republicans and of *acting* as citizens."[25]

In contemporary political science, low voter turn out, citizen apathy, the triviality of political campaigns are often cited as consequences of the failure of modern democracies to include citizens in meaningful activities. Much of the recent discussion among social scientists about "participatory democracy" and "strong democracy" speculates about ways to remedy these shortcomings.[26] But other than noticing the pungent effects of television upon election campaigns and the pervasive effects of modern consumerism, social scientists seldom take note of the connection between the hollowness of modern citizenship and the social relations of technology.

In fact, the political vacuum evident in the lack of citizen roles, citizen awareness, and citizen speech within liberal democratic society is greatly magnified within today's technology-centered workplace. Devices and systems commonly used in factories, fields, shops, and offices seek productivity and profit by controlling human behavior. In such settings the spontaneity and variability of workers' activities are regarded as a cause of uncertainty and a risk for business. For that reason the physical movements and decision-making abilities of employees are subject to rational planning and centralized guidance. Rather than encourage personal autonomy, creativity, and moral responsibility, many jobs and machines are designed to eliminate these qualities altogether.[27]

One might suppose that the technical professions offer greater latitude in dealing with the moral and political dimensions of technological choice. Indeed, the codes of engineering societies mention the higher purposes of serving humanity and the public good, while universities often offer special ethics courses for students majoring in science and engineering.[28] As a practical matter, however, the moral autonomy of engineering and other technical professionals is highly circumscribed. The historical evolution of modern engineering has placed most practitioners within business firms and government agencies where loyalty to the ends of the organization is paramount. During the 1920s and 1930s there were serious attempts to change this pattern, to organize the various fields of engineering as truly independent professions similar to medicine and law, attempts sometimes justified as ways to achieve more responsible control of emerging technologies. These efforts, however, were undermined by the opposition of business interests that worked to establish company loyalty as the engineer's central moral concern.[29] Calls for a higher degree of "ethical responsibility" among engineers are still heard in courses in technical universities and in obligatory after-dinner speeches at engineering societies. But pleas of this sort remain largely disingenuous, for there are few legitimate roles or organized settings in which such responsibility can be strongly expressed.

One could expand the inventory of social vocations in which moral issues in technological choice might be deliberated and decided, to include business managers, public officials, and the citizenry at large. Alas, there is little evidence that anything about these roles adds qualities of ethical reflection or action missing in ordinary workers or technical professionals. The responsibility of business managers is to maintain the profitability of the firm, a posture that usually excludes attention to the ethics of technological choice. Where questions of responsibility arise, businessmen usually listen to hired lawyers who explain their legal liabilities. Elected officials, similarly, find little occasion to consider the moral dimensions of technological choices. Their standard approach is to consult the opinions of scientific and technical experts, judging this information in ways that reflect a variety of economic and political interests. The general public may have a vague awareness of policy choices in energy, transportation, biomedical technology, and the like. But its response is increasingly apathetic, reactive, and video-centered.

Under such circumstances it is not surprising to find that people who call for moral deliberation about specific technological choices find themselves isolated and beleaguered, working outside or even in defiance of established channels of power and authority. At the level of individual action one finds the hero of much contemporary writing about technology and ethics—the "whistleblower," an employee who notices something troubling in the day-to-day workings of a sociotechnical system and tries to call it to the attention of a reluctant employer or the news media. By all accounts, such behavior is often severely punished by the organizations whose actions and policies the whistleblowers criticize. When they cannot be simply ignored, whistleblowers are isolated, fired from their jobs, and then black-balled within their professions. Their lives become embroiled in exhausting efforts to show the truth of their claims and reestablish their value as employees.[30] For career-minded students who study the stories of whistleblowers in university ethics courses, the underlying message is (regardless of what their teachers may intend): this is what happens if you speak out.

At the level of collective social action the method commonly used for expressing moral concerns about technological matters is that of "public interest" or "citizens" groups. Organized around key issues of the day, such groups take it upon themselves to express the interests and concerns of an otherwise silent populace about such matters as the arms race, nuclear power, environmental degradation, abortion, and many other issues. Ralph Nader, Helen Caldicott, and Jeremy Rifkin are among the contemporary figures who have become skillful in using this persuasive approach. It is characteristic of interest groups of this kind to be external to established, authoritative channels of decision-making power. The explicit purpose of groups identifying themselves with the "public interest" and "social responsibility" is to apply pressure, external pressure, upon political processes that otherwise move in what group members see as undesirable directions.

While the activities of public interest groups are clearly an exercise of the right of free speech, and while they are obviously important to the effective operation of modern democracy, the very existence of these groups points to the lack of any clear, substantive meaning for the term *public*. In this conception, the "public" arises *ad hoc* around certain points of social stress. One can claim to speak for "the public" simply by staging a demonstration or appearing on morning television news programs. The ease with which activists appropriate the word *public* leads to charges that particular groups are, in fact, unrepresentative, that "they don't represent my idea of the public interest." Nevertheless, public interest organizations offer the most direct means liberal democracies now have for focusing and mobilizing the concerns of ordinary people about controversial technologies.

The lack of any coherent identity for the "public" or of well-organized, legitimate channels for public participation contributes to two distinctive features of contemporary policy debates about technology, (1) futile rituals of expert advice and (2) interminable disagreements about which choices are morally justified.

Disputes about technology policy often arise in topic areas that seem to require years of training in fields of highly esoteric, science-based knowledge. A widely accepted notion about science is that it offers a precise, objective understanding of the world. Because technology is regarded as "applied science," and because the consequences of these applications involve such matters as complicated scientific measurements and the interpretation of arcane data, a common response is to turn to experts and expert research findings in hope of settling key policy questions.

This faith in scientific and technical advice involves much frustration in actual practice. Often it turns out that deep-seated uncertainties cannot be dispelled by consulting the experts. For the search for an objective answer brings a plurality of responses rather than a simple consensus. Studying the probable effects of background radiation, for example, different fields of scientific research give very different estimates of possible hazards. Problems of this kind are compounded by the fact that expertise is often indelibly linked to and biased by particular social interests. For example, looking at the problem of toxic waste disposal at Love Canal near Niagara Falls, New York, in the late 1970s, different social interests proposed different scientific models of the boundaries of the question and produced drastically different estimates of the hazards to citizens living in the area.[31] If, as contemporary sociologists claim, scientific knowledge is socially constructed, then scientific findings used in policy deliberations are doubly so. To an increasing extent, lawmakers and bureaucrats see scientific studies merely as resources to be deployed in ongoing power struggles.

What this suggests is that political disputes about technology are seldom if ever settled by calling upon the advice of experts. At public hearings held before legislative bodies, different social interests parade carefully chosen scientists and technical professionals. All of them speak with a confident air of "objectivity," but the experts often do not agree. Even where there is agreement about the "facts," there are still bound to be disagreements about how the "facts" are to be interpreted or what action is appropriate as a consequence.

Another characteristic of contemporary discussions about technology policy is that, as Alasdair MacIntyre might have predicted, they involve what seem to be interminable moral controversies. In a typical dispute, one side offers policy proposals based upon what seem to be ethically sound moral arguments. Then the opposing side urges entirely different policies using arguments that appear equally well-grounded. The likelihood that the two (or more) sides can locate common ground is virtually nil. Consider the following arguments, ones fairly typical of today's technology policy debates.

1a. Conditions of international competitiveness require measures to reduce production costs. Automation realized through the computerization of office and factory work is clearly the best way to do this at present. Even though it involves eliminating jobs, rapid automation is the way to achieve the greatest good for the greatest number in advanced industrial society.

b. The strength of any economy depends upon the skills of people who actually do the work. Skills of this kind arise from traditions of practice handed down from one generation to the next. Automation that deskills the work process ought to be rejected because it undermines the well-being of workers and harms their ability to contribute to society.

2a. A great many technologies involve risks of one kind or another. Judging the risks of chemical pesticides, one must balance the social benefits they bring against the risks they pose to human health and the environment. Considering the whole spectrum of benefits and risks involved, the good in using pesticides far outweighs their possible dangers.

b. Persons have a right to be protected from harm, including possible harm that may stem from useful technological applications. The use of pesticides subjects consumers to health hazards over which they have

little or no control. Regardless of the larger good that the use of pesticides might bring, their use should be curtailed to prevent the risk of
harm to individual consumers.

Positions of this kind involve a mixture of what may be highly uncertain empirical claims combined with philosophical arguments about which there is little consensus. Parties who square off in disputes of this kind usually believe that their side draws upon the very best data available and strong moral principles as well. But as the combatants circle each other in the ring, there is often a gnawing feeling that the various lines of moral reasoning have been concocted on the spot, used to justify positions that could be better described as emotional judgments or matters of sheer self-interest. In this way debates about technology policy confirm MacIntyre's argument that modern societies lack the kinds of coherent social practice that might provide firm foundations for moral judgments and public policies.[32]

What usually happens in such cases is a process of "muddling through." Interest groups apply pressure on politicians, gaining influence in proportion to the amount of money a group has to spend on the effort. Lawsuits are filed on one side or the other or both. Lawyers and judges sort through the flagrantly one-sided legal briefs, seeking precedents that might be patched together to provide a framework for deciding the case at hand. Television ads bombard viewers with flashy images and ten-second "sound bytes." Public opinion polls monitor the level of support for various proposals. Candidates for election sometimes take stands on issues that can then be included among the influences that sway voters in one direction or another. Eventually a policy outcome of some kind evolves, but it is seldom one that contains any experience of social learning that might be applied to similar episodes in the future.

Redefining Citizenship

In summary, I have argued that as moral philosophy confronts contemporary technology-related issues, it does so in an intellectual and social vacuum, one located in a deep gap between the technical and political spheres established by both ancient and modern philosophers. I have pointed to some of the consequences of this situation for thinking about technological choices and technology policies in our time. From this point of view, the technocratic approach I mentioned earlier—rushing forward with philosophical expertise to clarify moral categories, theories, and arguments in the hope that policymakers or the public will find them decisive—is a forlorn strategy. For the trouble is not that we lack good arguments and theories, but rather that modern politics simply does not provide appropriate roles and institutions in which the goal of defining the common good in technology policy is a legitimate project.

Under these circumstances a more fruitful path for philosophy is to begin exploring ways in which publics suited to renewed discussion about technological choices and policies might be constituted. Rather than echo the judgments of Aristotle and Adam Smith that political and technical affairs are essentially different, contemporary philosophers need to examine that question anew.

Some interesting possibilities arise in the fact that at long last the conceptual and practical boundaries between technology and politics upheld in both ancient and modern theory have begun to collapse. In the world of the late twentieth century, the spheres of technical and political life have merged in a variety of ways, woven together in situations in which common forms of human living have become dependent upon and shaped by technological devices and systems in telecommunications, computing, medicine, mass production, transportation, agriculture, and the like. To an increasing extent the qualities of technical artifacts reflect the possibilities of human living, what human beings are and aspire to be. At the same time, people mirror the technologies which surround them. Each day we see a widening of the kinds of human activities and consciousness that are technically embedded and technically mediated.

Although this rapidly growing, planetary technopolis strongly influences what our lives contain, few have tried to imagine forms of citizenship appropriate to this way of being. Some observers are content to point out the obvious, namely that technology is already highly politicized, that the development, introduction, and use of technologies of various kinds are always shaped by conflicts, negotiations, and machinations among powerful social interests. But to notice this fact is by no means to acknowledge the technopolitical sphere as a public space where citizen deliberation and action ought to be encouraged. To take that step, one must move beyond supposedly neutral sociological descriptions and explanations of how technologies arise and begin raising questions about the proper relationship between democratic citizenship and the shaping of technological order.[33]

Attempts of this kind have been launched recently in several modest experiments within the Scandinavian social democracies. These experiments are interesting in their own right, but also show the promise of creating citizen roles in places where private calculations of efficiency and effectiveness, costs, risks, benefits, and profits usually rule the day. A prototype of this variety of technological citizenship took shape at a research institute in Stockholm, the Center for Working Life. The basic goal of the Center's work was to expand the scope of Scandinavian ideals of worker democracy in which technological innovation was likely to occur. They were encouraged by Swedish laws passed in the middle 1970s that recognized the right of all parties in the workplace, managers and workers alike, to negotiate about matters that affect the quality of working life. The "co-determination laws" cover such areas as job allocation, training, and work environment. Beginning in the 1970s, legal rights of this kind were carried in a novel direction by a group of labor unions working with university-educated computer scientists and systems designers. Realizing that computerization was likely to transform Swedish factories, shops, and offices, fearing the loss of jobs and workers' skills, the teams set out to investigate the new technologies and to explore possible alternatives.[34]

In one such case, the UTOPIA project of the early 1980s, workers in the Swedish newspaper industry—typesetters, lithographers, graphic artists, and the like—joined with representatives from management and with university computer scientists to design a new system of computerized graphics used in newspaper layout and typesetting. The first phase of the project was to survey existing work practices, techniques, and training in the graphic industries. The group then formed a design workshop to consider possibilities for a new system, using a paper-and-plywood mock-up as the model of a newspaper workstation. From there they produced a forty-eight-page technical document giving precise design specifications to the computer suppliers.

The pilot system, installed at the Stockholm daily newspaper *Aftonbladet*, offers a pattern of hardware, software, and human relationship very different from what would have been produced by managers and engineers alone. It allows graphics workers considerable latitude in arranging texts and images, retaining many of their traditional skills, but realizing them in a computerized form. In their deliberations, project members considered but rejected the pre-packed graphics programs promoted by vendors from the United States because they reflected an "anti-democratic and de-skilling approach."[35] As project member and computer scientist Pelle Ehn observes, "What was new was that these technical requirements were derived from the principle that the equipment should serve *as tools for skilled work* and for production of *good use quality products*."[36]

The "Scandinavian approach" to participation in design is interesting not only for its tangible results but also for what it suggests about a positive politics of technology seen in broader perspective. In a small and tentative manner, the UTOPIA project created a public space for the political deliberation about the qualities of an emerging technical artifact. A diverse set of needs, viewpoints, and priorities came together to determine which material and social patterns would be designed, built, and put into operation. As Pelle Ehn points out, the important step in this process was to find a "project language game" in which all the participants from very different vocations, professions, and social backgrounds could speak to each other.[37] True, it was a fairly limited public that was constituted here. But it was far more inclusive than is normally the case in the printing industry or elsewhere.[38]

The creation of public spaces of this kind is, of course, predicated on modifying the right of owners of private property to have exclusive or even primary control of the shape of new technologies that affect how others live. That condition is, to a great extent, an accomplishment peculiar to Scandinavian social democracy, a product of political conflicts and agreements over the past several decades. It is now a condition sustained by the fact that more than 80 percent of Swedish workers are union members.[39]

Another achievement of the "Scandinavian approach" is to eliminate what I noted earlier as one of the most troubling features in contemporary technology policy: the ritual of expertise. In the UTOPIA project and others similar to it, a person's initial lack of knowledge of a domain of complex technical knowledge does not create a barrier to participation. The information and ideas needed to participate are mastered as part of a process in which the equality of team members is the established norm. Working from the opposite direction, those who came to the process with university degrees and professional qualifications explicitly rejected the idea that they were the desig-nated, authoritative problem-solvers. Instead they offered themselves as persons whose knowledge of computers and systems design could contribute to discussions conducted in democratic ways.

This approach may also help dispel the second disturbing feature of contemporary technology policy debates, the interminable moral controversies they tend to generate. Here the guiding assumption is that if people with diverse viewpoints and conflicting social interests come together as equals in a situation that presents a common problem to be solved, an agreement will eventually evolve. As Ehn describes a typical predica-ment, "Management introduces new technology to save manpower. Journalists, graph-ics workers, and administrative staff confront each other in the struggle over a decreas-ing number of jobs. Is there a basis for solving these demarcation disputes across professional and union-based frontiers? Can a new way of organizing work create peaceful coexistence in the borderland?"[40] The answer seems to be yes. However, the answer is never as simple as one set of philosophically well-grounded prescriptions winning out over another. Instead what happens is a negotiated political agreement among those whose interests will be affected by the change.

What the Scandinavian projects have done in an experimental way is to institute technopolitical practices from which new citizen virtues call emerge. Within small communities constituted for the purpose, choices about technologies that will influ-ence the quality of social life are carefully studied and debated. This involves no ex-pectations of political heroism, only the sense that ordinary people, regardless of back-ground or prior expertise, are capable of taking a turn making decisions of this kind.[41] The vision of knowledge and social policy that underlies these efforts strongly resem-bles Paul Feyerabend's anarchistic proposals for "committees of laymen" involved in science.[42] In this instance, however, there was an opportunity to test the ideas in actual practice.

As revealed by Ehn's engaging treatise *Work-Oriented Design of Computer Arti-facts*, the role of philosophy in this process is a limited but useful one. It attempts to clarify the basic conditions that undergird practices of work and discourse within the design projects. By seeking to understand these practices at a deeper, more general level, philosophical inquiry may shed light on ongoing negotiations as they occur. Thus, Ehn draws upon the writings of Heidegger, Wittgenstein, Habermas, and other philosophers to illuminate his central concerns.[43] In the ideal case, philosophical reflection becomes one element in the process, although not one given privileged status. For it is understood that the key insights, lessons, and prescriptions must arise from a process in which project members, regarded as equals, join to explore the properties of both technical artifacts and social arrangements in a variety of configurations.

A criticism that might be raised about approaches like that pursued by Ehn and his Scandinavian colleagues is that they work at a superficial level within the technologies they confront. As the historian of technology Ulrich Wengenroth has noted, there is today a widening gap between "professionalization" and "trivialization" in many fields of technological development. Deeper, more complex levels of technical design and operation—the making of computer chips, for example—are accessible to and acted upon by only a handful of technical professionals. The same technologies are, however, restructured at the level of the user interface and present themselves in a deceptively friendly form. As Wegenroth observes, "If a new technology is met by suspicion and resistance in society, its acceptance is not won by reducing its complexity to make it intelligible and thus controllable by the general public, but by reengineering its interface to trivialize it."[44]

Do the Scandinavian projects merely retailor interfaces to make them more agreeable to workers while leaving the deeper structures of the technology as something given? The question cannot be answered in this brief overview. It is worth noting, however, that within the domain of computer programming the innovations of the Scandinavian researchers appear to be fairly deep-seeking. As noted, members of the UTOPIA project rejected an American firm's software package because it contained entrenched forms of hierarchical work organization, features that the group found "anti-democratic and de-skilling." Rather than try to weed out the deep-seated authoritarianism of American computer programs, the UTOPIA project elected to start from scratch.[45]

It is perhaps too early to characterize the virtues of citizen participation that might emerge from practices of this kind, too soon to specify whether this experience might be successfully applied to realms of technological choice usually governed by the merciless logic of economic and technical rationalization.[46] Members of the UTOPIA project appear to have developed a sense of cooperation, caution, and concern for the justice of their decisions. They were especially conscientious in trying to find effective designs that could take advantage of computer power while preserving the qualities of traditional workmanship. The members realized that conditions expressed in the design of a new system were conditions they would eventually have to live with. In that way their work echoes Aristotle's definition of the virtue of the good citizen, namely an understanding of both how to rule and be ruled. At a time in which politics and technology are thoroughly interwoven, perhaps a similar definition of the virtue of citizens is that they know both how to participate in the shaping of technologies of various kinds and how to accept the shaping force that these technologies will eventually impose.

From this viewpoint the creation of arenas for the politics of technological choice is much more than a way of solving unsettling problems that arise in the course of technological change, although steps of this kind certainly might do that. It is also more than finding alternatives to the increasingly absurd logic of efficiency, productivity, and control that now drives technological choices in the global economy, although there is certainly a need for such alternatives. Even more important, the creation of new spaces and roles for technological choice might lead us to affirm a missing feature in modern citizenship: the freedom experienced in communities where making things and taking action are one and the same.

Notes and References

1. Robert Sinsheimer, "Genetic Engineering: Life as a Plaything," in A. Pablo Iannone (ed.), *Contemporary Moral Controversies in Technology* (New York: Oxford University Press, 1987), p. 131.

2. James H. Moor, "What Is Computer Ethics?," *Metaphilosophy* 16 (1985), no. 4, p. 275.

3. My treatment of classic and modern attitudes toward technology draws upon Carl Mitcham's excellent survey, "Three Ways of Being-With Technology," in Gayle Ormiston (ed.), *From Artifact to Habitat: Studies in the Critical Engagement of Technology* (Bethlehem, PA: Lehigh University Press, 1990).

4. As Plato explains in *The Republic*, the real table is not that made by a craftsman, but the table that exists as an ideal form in the transcendent realm. Attempts to define the good society must understand this, seeking true rather than debased foundations for political practice. For that reason, Plato places the arts and crafts in the lowest of three social classes, and removes from them any chance of holding power. While he recognizes that agriculture, medicine, architecture, and the other practical arts are necessary to the life of the state, they offer nothing of value in ruling a good society. In both *The Republic* and *The Laws*, Plato advises those who would rule to stay as far away from mundane technical activities as possible. See my discussion of Plato's views in *The Whale and the Reactor: A Search for Limits in an Age of High Technology* (Chicago: University of Chicago Press, 1986), chapter 3.

5. Plato, *The Laws of Plato*, trans. Thomas L. Pangle (Chicago: University of Chicago Press, 1980), 797d.

6. Lucretius, *The Nature of Things*, trans. Ronald Latham (Baltimore: Penguin Books, 1951), p. 211.

7. Aristotle, *Politics*, trans. Benjamin Jowett in Jonathan Barnes (ed.), *The Complete Works of Aristotle*, vol. II (Princeton: Princeton University Press, 1984), p. 1987.

8. Ibid., p. 2026.

9. Ibid., p. 2027.

10. Ibid., pp. 2028f.

11. Ibid., p. 2027.

12. M. I. Finley, *The Ancient Economy* (Berkeley, CA: University of California Press, 1973), pp. 75–76.

13. Ibid., p. 75.

14. Ibid., p. 84.

15. Francis Bacon, *The Great Instauration*, in Fulton Anderson (ed.), *The New Organon and Related Writings* (Indianapolis: Bobbs-Merrill Co., 1960), p. 8.

16. Ibid., p. 117.

17. Jean Le Rond D'Alembert, *Preliminary Discourse to the Encyclopedia of Diderot*, trans. Richard N. Schwab (Indianapolis: Bobbs-Merrill, 1963), p. 42.

18. See Albert O. Hirschman, *The Passions and the Interests: Political Arguments for Capitalism before Its Triumph* (Princeton: Princeton University Press, 1977).

19. Baron de Montesquieu, *The Spirit of Laws*, trans. Thomas Nugent, rev. ed., vol. I (New York: P. F. Collier & Son, 1900), p. 46.

20. C. B. MacPherson, *The Theory of Possessive Individualism: Hobbes to Locke* (Oxford: Clarendon Press, 1962).

21. David Hume, "Of Refinements in the Arts," in T. H. Green and T. H. Grouse (eds.), *The Philosophical Works*, vol. 3, reprint of new ed. of 1882 (Aalen: Scientific Verlag, 1964), p. 301.

22. Adam Smith, *The Wealth of Nations*, bks I–III with introduction by Andrew Skinner (Harmondsworth: Penguin, 1970), p. 446.

23. Ibid., p. 443.

24. Ibid., p. 446.

25. Hannah Arendt, *On Revolution* (Harmondsworth: Penguin Books, 1977), p. 253.

26. See, e.g., Benjamin Barber, *Strong Democracy: Participatory Politics for a New Age* (Berkeley, CA: University of California Press, 1984).

27. For poignant descriptions of circumstances that often face workers, see Barbara Garson, *Electronic Sweatshop: How Computers Are Transforming the Office of the Future into the Factory of the Past* (New York: Simon & Schuster, 1988).

28. See, e.g., Peter Windt et al. (eds.), *Ethical Issues in the Professions* (Englewood Cliffs, N.J.: Prentice Hall, 1989), and Deborah G. Johnson (ed.), *Ethical Issues in Engineering* (Englewood Cliffs, N.J.: Prentice-Hall, 1991). My essay "Engineering Ethics and Political Imagination," in Paul T. Durbin (ed.), *Broad and Narrow Interpretations of Philosophy of Technology* (Dordrecht: Kluwer Academic Publishers, 1990), pp. 53–64, criticizes the approaches often used to teach ethics for technical professionals.

29. Edwin Layton, *Revolt of the Engineers: Social Responsibility and the American Engineering Profession* (Cleveland: Case Western Reserve University, 1971), chapters 1–2.

30. Myron Glazer and Penina Glazer, *The Whistleblowers: Exposing Corruption in Government and Industry* (New York: Basic Books, 1989).

31. Beth Savan, *Science under Siege: The Myth of Objectivity in Scientific Research* (Montreal: CBC Enterprises, 1988).

32. Alasdair MacIntyre, *After Virtue: A Study in Moral Theory*, 2nd ed. (Notre Dame: University of Notre Dame Press, 1984), chapters 14 and 15.

33. For a critique of the new sociology of technology, see my "Social Constructivism: Opening the Black Box and Finding It Empty," *Science as Culture*, no. 16 (Autumn 1992).

34. For a description of Scandinavian experiments in democratic participation in design, see Pelle Ehn, *Work-Oriented Design of Computer Artifacts* (Stockholm: Arbetslivcentrum, 1988).

35. Ibid., p. 345.

36. Ibid., p. 339 (italics in the original text).

37. Ibid, p. 17.

38. In fact, problems arose within the UTOPIA project because it was not inclusive enough, excluding the participation of journalists. As Ehn notes, the future of the project "depends upon whether the graphic workers and journalists succeed in overcoming their professional clash of interests, and together develop a common strategy." Ehn, op. cit., p. 357.

39. Peter Lawrence and Tony Spybey, *Management and Society in Sweden* (London: Routledge & Kegan Paul, 1986), p. 85. For an overview of the relationship between technology and work in Sweden, see Åke Sandberg, *Technological Change and Co-Determination in Sweden: Background and Analysis of Trade Union and Managerial Strategies* (Philadelphia: Temple University Press, 1992). An excellent discussion of the moral issues confronting Scandinavian social democ-

racy can be found in Alan Wolfe, *Whose Keeper?: Social Science and Moral Obligation* (Berkeley, CA: University of California Press, 1989).

40. Ehn, op. cit., p. 342.

41. For a general exploration of tensions between technical expertise and direct democracy, see Langdon Winner (ed.), *Democracy in a Technological Society* (Dordrecht: Kluwer Academic Publishers, 1992), and Frank Fischer, *Technocracy and the Politics of Expertise* (Sage Publications, Newbury Park, CA, 1990).

42. See Paul K. Feyerabend, *Science in a Free Society* (London: NLB, 1978), and his suggestions in "Democracy, Elitism, and Scientific Method," *Inquiry* 23(1) (1980), pp. 3–18.

43. Arguments and conclusions similar to Pelle Ehn's can be found in Terry Winograd and Fernando Flores, *Understanding Computers and Cognition* (Reading, MA: Addison-Wesley, 1987).

44. Ulrich Wengenroth, "The Cultural Bearings of Modern Technological Development," in Francis Sejersted and Ingunn Moser (eds.), *Humanistic Perspectives on Technology, Development and Environment* (Oslo: Centre for Technology and Culture, Report Series No. 3, 1992).

45. Ehn, op. cit., pp. 344–45.

46. Methods of organizing people and machinery in the mode of "just-in-time" and "lean production," now gaining momentum in the global market economy, point in directions much different from those pursued by Scandinavian workplace reformers. The workplace regimes created within this mode of production could well achieve levels of rationalization and centralization that would make Frederick W. Taylor and Jacques Ellul blush. See J. P. Womack, D. T. Jones and D. Roos, *The Machine that Changed the World* (New York: Rawson Associates, 1990).

5

The Moral Significance of the Material Culture

Albert Borgmann

M ODERN PHILOSOPHY HAS been at two removes from the real world. First, in aspiring to theory, it has been distanced from practice. Theory can inform practice, but practice is richer than theory and, above all, self-sustaining. Practice can survive without theory while theory arises from a practice and perishes without the nourishment of a practice. Practice, as philosophers have always seen it, is in turn removed from its tangible setting. Yet material culture constrains and details practice decisively. Practice, abstracted from its tangible circumstances, is reduced to gesturing and sometimes to posturing.

Philosophy as we know it began with Plato, and in the beginning material reality was thought to be the adversary and seducer of philosophy. To philosophize was to rise above the tangible phenomena to the intelligible ideas. And while Aristotle acknowledged the life of pleasure and the life of honor and action, it is the life of contemplation that constitutes human fulfillment. Contemplation in Greek is *theoria*; with Aristotle the word and the vision that were to rule philosophy came to the fore. They continued their reign through the Middle Ages where the *vita contemplativa* was considered superior to the *vita activa*.

Practice, to be sure, was never far from ancient and medieval theory. To know the good is to do the good, says Plato. Virtue, says Aristotle, is a skilled practice. The 119 metaphysical questions of Thomas Aquinas's *Summa Theologica* are followed by 303 questions on ethics, 189 of them on virtues. Practice, in turn, overshadowed tangible reality. Why? Practice, for the ancients and medievals, was enacted on a solid and familiar stage. Nature presented the powerful and regular backdrop of human life. Material culture presented a similarly firm and surveyable precinct. Where it changed, it did so, within any two or three generations, slowly and only in part. Not that the ancients and medievals were entirely unconcerned about the material world. They worried that it might provoke recklessness in the way humans shape it to their purposes and extravagance in the way they enjoy it. But all in all they took the material culture to be so solid and familiar that its direct bearing on philosophy could be handled in parentheses and asides.[1] Where moral virtues, chiefly temperance, referred to material goods, the solidity and familiarity of the latter authorized philosophers simply to presuppose them and to concentrate attention on the former.

Philosophical theory underwent a radical transformation at the beginning of the modern era. The most consequential development was the emigration and emancipation of the theory of nature from philosophy, a development that resulted in modern science.[2] From the start, beginning with Descartes and Bacon, natural science had for one of its objects the transformation of material nature for the liberation and enrichment of human life. The realization of this program was slow in coming, but it had an overpowering effect on philosophy from the very beginning of the modern period. The effect was twofold. For one thing, philosophers without question accepted the scientific characterization of the fundamental transformation of material culture that was to result from scientific research. It would be the fulfillment of a promise of liberty and prosperity. Who would think of advocating servitude and poverty instead? For another, philosophers assimilated their enterprise to science but sought to execute it at a higher level. If science was to furnish a theory that would lead to an enlightened reconstruction of the physical environment, philosophy would provide a theory that was to bring about a salutary reconstruction of reality entire, including not only science and its subject, but also art, religion, politics, knowledge, and human conduct.

This then was the course of events that, in the modern period, removed philosophy from human practice and material culture. The latter was surrendered to science for illumination and transformation. The former was subjected to the reconstructive ambition of philosophical theory and ceased to be a subject of reflection in its own right.

While science by way of technology realized a certain version of its epochal project, the ambitions of philosophical theory became ever more marginal to modern society.[3] It was the pretentious poverty of theory that led Heidegger in 1927 and Oakeshott in the 1950s to reconsider and rehabilitate the dignity of practice.[4] Meanwhile feminism and postmodern particularism have attacked universal theory on a broad front and have delineated, elucidated, and recounted a great variety of human practices. All this is part of a commendable endeavor to recover the richness of life, to acknowledge the fragility and contingency of human circumstances, and to encourage the gentler voices in the conversation of humanity.

But if at its postmodern dawn philosophy has returned to practice, it has remained distant from material culture. The ontological enterprise, i.e., the endeavor to provide a philosophical theory that would illuminate the dark regions and clarify the confusions of all of reality, including material reality, has been stalemated.[5] Aesthetics does attend to the material embodiment of art, and it has begun its journey back to the fork where it parted ways with ethics.[6] Still, these efforts are too limited and tentative to cast much light on the moral charge of material culture. A similar limitation holds for the close historical studies of material culture one finds in journals such as *Technology and Culture* and *Material Culture*. If there is an implicit ethics in such studies, it is still the Enlightenment morality that vaguely applauds liberty, prosperity, and diversity.

Liberal and Marxist political philosophy has been keenly alert to the ethical significance of the distribution of material goods. But all color, texture, and flavor have been bleached from material culture. It has been reduced to one aspect, viz. power. The distribution of goods is taken to be crucially and finally a distribution of power. The commendable goal of leftist analysis is equality or at least more of it. But the poverty of reducing material culture to power relations comes to the fore when Marx or Rawls contemplate the human condition once (greater) equality is achieved. What remains as the challenge of the good life is a bland and insubstantial notion of self-realization.[7]

A similar reductionism emerges when anthropologist Mary Douglas considers *The World of Goods.*[8] Goods are reduced to information of which people possess more or less. They use it to communicate to one another their social standing; and throughout the different social ranks, power once more is the spine of the body politic.[9] Still, Douglas belongs to a loosely distinguishable group of social theorists and cultural critics who have been struck by the novel and troubling character of the material culture in the advanced industrial countries and have illuminated this issue in various ways.[10] But none dares to proceed to an explicit moral assessment. That is left to the professional moralists, the philosophers. Thus the moral significance of the material culture falls between the stools of the professions.

In at least one case, however, a sociological analysis of an important segment of our material setting, the home, leads right up to and some way into a moral examination. Equally important, sociologists Mihaly Csikszentmihalyi and Eugene Rochberg-Halton remark expressly on how little is known about *The Meaning of Things.*[11] Lifting the veil of inconsiderability, the authors uncover what I take to be the crucial issue in the study of material culture.

It is a distinction between two kinds of things or reality. The first constitutes globally "a universe that speaks to humans."[12] Locally it comes to the fore in any object that conveys "meaning through its own inherent qualities" and through "the active contribution of the thing itself to the meaning process."[13] It is a kind of reality, however, that the authors find to be rarely acknowledged in their extensive interviews. Things are largely treated as semantically pliable material whose significance is shaped through an investment of psychic energy. There is much leeway for the emotional and moral shape a certain thing can be given.[14]

Let me call these two kinds of things and reality commanding and disposable. The latter kind is not unstructured, of course. We know as much from our command of commonsense distinctions. Nor is it intuitively obvious or trivial what kinds of things people especially endow with value in their homes.[15] But for the purposes of understanding the moral salience of the material culture, Csikszentmihalyi and Rochberg-Halton's key finding is as important as it is disappointing—of itself material culture does not seem to matter much one way or another. Although, to take one example, they find strikingly different syndromes of goals, participation in the public sphere, the choice of role models, and personality patterns for "warm" as opposed to "cool" homes, "the kinds of objects mentioned by the two groups were essentially the same. . . . What did separate the two groups was simply the interpersonal meaning associated with the objects."[16]

Yet this disappointment rests on the assumption that the eclipse of commanding reality and the prominence of disposable reality as the normal arena of human conduct is itself morally inconsiderable. To raise the issue, consider music as an instance where the decline of commanding and the prominence of disposable reality comes into focus. In Csikszentmihalyi and Rochberg-Halton's study, music is considered under the headings of musical instruments and stereos.

A traditional musical instrument is surely a commanding thing. It is such simply as a physical entity, finely crafted of wood or metal, embodying centuries of development and refinement, sometimes showing the very traces of its service to many generations. An instrument particularly commands the attention of the student who, unless she is a prodigy, must through endless and painstaking practice adjust her body to the exacting requirements of this eminently sensitive thing. She must, moreover, train her eyes to read the music and her mind to transpose the visual information rapidly and easily into those bodily maneuvers the instrument will register. Music performed on an instrument in our presence captures our attention, and not only our ears but our eyes as well. We see the strings vibrate, the wood resonate, and the metal shape the sounding air. And the instrument, while it produces sound, reveals a person too, the grace or the strain or the fervor of the one who plays the instrument.

A stereo produces music as well or, in fact, much better, i.e., with the supernatural sonority and consistency that no live performance can sustain and with a range in the kinds of music that would require a standing army of virtuosi and virtuosae were it to be humanly available to a listener's call and beckon. Is a stereo a commanding piece of reality? As a physical entity, some sets are imposing in their size and high-tech gleam. Yet some stereos are praised for their slender and self-effacing physical appearance. As a thing to be operated, a stereo is certainly not demanding. Nor do we feel indebted to its presence the way we do when we listen to a musician. We respect a musician, we own a stereo. A set and the music it produces are entirely at our disposal.

In Csikszentmihalyi and Rochberg-Halton's research, instruments are mentioned as often as stereos among the objects people most cherish in their homes. Stereos and instruments rank fifth and sixth after furniture, visual art, photographs, and books.[17] But to cherish an object is one thing, to use it another. It is apparent from Csikszentmihalyi and Rochberg-Halton's discussion that people cherish stereos because they use and depend on them while an instrument is often valued as "a central symbol of a lifestyle once cherished or anticipated in the future."[18] Studies of how people in this country typically spend their time confirm these indications and our general intuition that musical instruments are rarely played in American homes, on the average three minutes or so a week, a seventh or eighth of the time that is spent on listening to recorded music.[19]

The case of music, then, instantiates the eclipse of commanding reality. It also affords the clues as to why and how this is happening, the why and the how being tightly connected. Why is recorded music preferred to performed music? The latter is arduous to master and limited in its range. But a complaint about such defects can become consequential only when there is a way of providing music effortlessly and abundantly. The history of the technology of recorded music is the history of obliging ever more fully the complaint about the burden and confinement of live music. Or to put the development more positively, at the beginning of the invention of recorded music stands a promise of disburdenment and enrichment, the promise to provide music freely and abundantly. This promise in turn is part of a larger one, viz. the promise of general liberty and prosperity—the promise that inaugurated the modern era.

In a democracy, such a promise cannot in principle be extended to a few and fulfilled through the servitude of the many. The task of furnishing a desirable good must be given over to a machinery. In 1876, Thomas A. Edison rigged up a cylinder covered with a metal foil on which a stylus, activated by the vibrations of a membrane, inscribed a wavy groove as the cylinder turned. The membrane in turn moved in response to the sound waves of voices or instruments. The grooved recording on the cylinder could be played back by turning the cylinder and having another stylus follow the groove and activate a membrane.[20] From this machine the contemporary stereo has descended through innumerable ingenious mutations.

An Edison phonograph has an engaging simplicity and intelligibility. The voice moves the membrane, the membrane moves the stylus, the stylus grooves the metal foil and leaves a record. Reversing the process we begin with the record and end with the sound. To be sure, it is hard for the layperson to appreciate that the richness and complexity of an orchestra can be reduced, for monaural purposes, to a single two-dimensional wave form and that ear and brain are capable of resolving that one compound waveform into the several components that represent the individual instruments. Yet the unintelligible residue of the Edison phonograph is limited and understandable as an articulate explanandum if not as regards its explanans.

The simplicity of the machinery corresponds of course to the crudeness of the commodity it produces. A simple phonograph produces poor sound. Correspondingly, today's stereo, which produces preternaturally perfect sound, is totally unintelligible to the typical consumer who does not even begin to understand the mathematics, logic, electronics, and mechanics that are embodied in a compact disc player and its associated equipment. As a consequence music has become a disembodied, freefloating something, a commodity that is instantly, ubiquitously, and easily available.

Music has been mechanized and commodified. These two processes are really one. Music can become available as a commodity only if there is a sophisticated and reliable machinery that will produce it at the consumer's will. We may call the conjunction of machinery and commodity a technological device. The stereo as a device contrasts with the instrument as a thing. A thing, in the sense in which I want to use the term, has an intelligible and accessible character and calls forth skilled and active human engagement. A thing requires practice while a device invites consumption. All this amounts to an explication of the distinction of realities derived from Csikszentmihalyi and Rochberg-Halton. Things constitute commanding reality, devices procure disposable reality.

If we inspect the typical American home from a historical perspective and with a view to its balance of things and devices, it becomes obvious that devices and consumption have replaced things and practices. Consider the culture of the table. The practice of cooking has been greatly diminished through the availability of convenience foods and microwave ovens. The practice of dining has been curtailed through grazing, snacking, and grabbing a bite to eat—forms of mere food consumption. The food itself has been reduced from a contextually intelligible and illuminating thing to an opaque if glamorous commodity.

Or consider the culture of the word, the traditional medium of world appropriation. People used to orient themselves in their world through the practice of writing letters, telling stories, engaging in conversations, attending plays, reading to one another and through the silent reading of books and newspapers. Much of the practice that used to animate and sustain the culture of the word is captured in the term *literacy*. The things that used to center these practices do not have the tangibility of instruments and foods. But they were things just the same, commanding and illuminating realities, tales, plays, and texts.

Telephone and television are the technological devices that have weakened literacy and impoverished the culture of the word. Electronic machines have disburdened us of the demands of reading and writing. Once we had to impart our worlds through the work of writing or telling, and we had to gather our worlds laboriously from the promptings of writing and our fund of experiences and recollections. Now information is handed to us as readily available sounds and sights. Engagement with the world has been yielding to the consumption of news and entertainment commodities.

If we move outside the home to consider the balance of things and devices and of practices and consumption in the public realm, we come to see not only that things have yielded to devices as they have in the home, but also that the machinery side of public devices is much more prominent than their commodity side. Consider the two most imposing public devices that the second half of this century has produced, the highway system and highrise buildings. They serve the productive and administrative machinery by providing for trucking, accounting, or lawyering. Or they serve private rather than public consumption as they do in furnishing personal transportation or housing. Settings for public consumption are relatively rare though they are prominent enough. Examples are Disneyland and Disney World, theme parks, and shopping malls.

The prevalence of the instrumental side in the public realm is mirrored in the activity that is typically devoted to it, viz. labor. We pay for the liberty and prosperity of (largely private) consumption through labor, the construction and maintenance of the technological machinery, and labor is done almost entirely in the public realm if largely in its private sector.

Assessing the moral significance of the material culture, then, comes in large part to asking what the moral consequences of the rule of the device paradigm are. Here again we can depart from the findings of Csikszentmihalyi and Rochberg-Halton's investigation of the attitudes and reactions that correspond to stereos and musical instruments. Music plays a stronger role among children and adolescents than among adults, but quite generally stereos are used by people individually "as a modulator of emotions, a way of compensating for negative feelings."[21] A musical instrument, to the contrary, condenses "a whole complex set of meanings." Consider Csikszentmihalyi and Rochberg-Halton's account of a baritone ukulele's significance in a man's life.

> It allows the man to use his skills in musical expressions, to have fun in the present while reliving past enjoyment, and at the same time, sharing the fun with those he loves. The ukulele in this case is a catalyst for a many-sided experience; it is not only an instrument for making sounds but is also a tool for a variety of pleasurable emotions. In playing it this man recaptures the past and binds his consciousness to that of others around him.[22]

Stereos appear to disengage people from their physical and social environment. When household objects are ranked according to the percentage of their meanings that respondents say refer to oneself, they rank second after television among eleven kinds of objects. Musical instruments rank fourth. When reference to others is the criterion, instruments rank fifth and stereos eighth.[23]

Considering other objects and criteria in Csikszentmihalyi and Rochberg-Halton's table, one finds, not surprisingly, that television, the only other technological device on the list, ranks even higher than stereos as a self-oriented object and as low as stereos in other-orientation. It is equally unsurprising if even more distressing that television and stereos rank first and second when it comes to meanings referring to experience (instruments ranking fourth). They rank last and third to last when reference is to memories (instruments rank seventh). One would also expect, as it turns out, that books, culturally most like instruments, rank very closely with the latter except in reference to others where books rank much lower.[24]

Such objective data provide abstract lineaments that need to be given color and concrete detail as well as a broader interpretation. Csikszentmihalyi and Rochberg-Halton provide the latter through a concentric conception of the self. The narrowest sphere is the individual and personal, inevitably embedded in a wider social sphere, and in need of an inclusive, cosmic orientation.[25] Every traditional culture shows some such structure. But Csikszentmihalyi and Rochberg-Halton were surprised to find at most "a glimmer of the cosmic self" in the ways people talked about their homes.[26] The balance of the personal and the cosmic self seems very much upset in favor of the personal.[27]

In sum, material culture in the advanced industrial democracies spans a spectrum from commanding to disposable reality. The former reality calls forth a life of engagement that is oriented within the physical and social world. The latter induces a life of distraction that is isolated from the environment and from other people. There are pairs of terms that detail further the styles of life corresponding to the endpoints of the cultural continuum, viz. excellent vs. banal, deep vs. shallow, communal vs. individualist, celebratory vs. consumerist, and others.

This vocabulary appears ambiguous or peremptory without its proper complement of material culture. Ethics, to be truly illuminating, must become real. Theoretical and practical ethics, when abstracted from reality, becomes and remains partial, i.e., both limited and unwittingly tendentious. Taking liberal democratic ethics as an example of theoretical ethics, we can see that its devotion to equality is in danger of being trivialized and subverted so long as what is finally to be distributed in equal shares remains debilitatingly banal. And its devotion to equality is becoming increasingly shrill and ineffective because liberal theorists overlook the narcotic effect that disposable reality has on people.

If we let virtue ethics with its various traditional and feminist variants stand in for practical ethics, we must recognize that virtue, thought of as a kind of skilled practice, cannot be neutral regarding its real setting. Just as the skill of reading animal tracks will not flourish in a metropolitan setting, so calls for the virtues of courage and care will remain inconsequential in a material culture designed to procure a comfortable and individualist life.

What would be the major consequence of complementing theoretical and practical ethics with real ethics? It would be the realization that we make our crucial decisions not as individuals, as consumers, taxpayers, and voters who navigate their course in preestablished and rigid channels, but as citizens, and not just as citizens who pass on matters of civil rights and social welfare, but as citizens who take responsibility for the large design of our culture and either make it hospitable to commanding reality or continue to thicken its suffocating overlay of disposable reality.

Putting matters this way is to clarify also what is at the heart of real ethics. It is not finally the desire for greater scholarly circumspection or radicality, but rather the readiness to answer to the claim of eloquent things and the concern to advance the kind of excellence that is the human counterpart to the splendor of reality.

Notes and References

1. Carl Mitcham, "Three Ways of Being-With Technology," in Gayle L. Ormiston (ed.), *From Artifact to Habitat* (Bethlehem, PA: Lehigh University Press, 1990), pp. 31–59.

2. Hans Jonas, "The Practical Uses of Theory," in *The Phenomenon of Life* (New York: Dell Publishing, 1966), pp. 188–210.

3. Richard Rorty, *Philosophy and the Mirror of Nature* (Princeton, NJ: Princeton University Press, 1979).

4. Martin Heidegger, *Sein und Zeit*, 9th ed. (Tübingen: Niemeyer, 1960), and Michael Oakeshott, *Rationalism in Politics* (New York: Basic Books, 1962).

5. Jaegwon Kim, "The Myth of Nonreductive Materialism," *Proceedings and Addresses of the American Philosophical Association* 63 (1989), pp. 31–47.

6. Suzi Gablik, *The Reenchantment of Art* (New York: Thames and Hudson, 1991).

7. Herbert Marcuse, *One-Dimensional Man* (Boston: Beacon, 1964), pp. 2, 16, 18, 235. John Rawls, *A Theory of Justice* (Cambridge, MA: Harvard University Press, 1971), pp. 424–33.

8. Mary Douglas and Baron Isherwood, *The World of Goods* (New York: Basic Books, 1979).

9. Ibid., see in particular p. 89.

10. E.g. August Heckscher, *The Public Happiness* (New York: Atheneum, 1962); Walter Kerr, *The Decline of Pleasure* (New York: Simon and Schuster, 1962); Staffan B. Linder, *The Harried Leisure Class* (New York: Columbia University Press, 1970); Daniel J. Boorstin, *Democracy and Its Discontents* (New York: Vintage Books, 1975); Tibor Scitovsky, *The Joyless Economy* (Oxford: Oxford University Press, 1976); Robert L. Heillbroner, *An Inquiry into the Human Prospect*, 2nd ed. (New York: Norton, 1980); Neil Postman, *Technopoly* (New York: Knopf, 1992).

11. Mihaly Csikszentmihalyi and Eugene Rochberg-Halton, *The Meaning of Things* (New York: Cambridge University Press, 1981), pp. 1 and 46–47.

12. Ibid., p. 12.

13. Ibid., p. 43.

14. Ibid., pp. 32, 65, 77, 87, 178.

15. Ibid., pp. 55–89.

16. Ibid., p. 165.

17. Ibid., p. 58.

18. Ibid., p. 73.

19. Martha S. Hill, "Patterns of Time Use," in F. Thomas Juster and Frank P. Stafford (eds.), *Time, Goods, and Well-Being* (Ann Arbor, MI: Survey Research Center, Institute for Social Research, University of Michigan, 1985), p. 173. These data from the mid-seventies lump music, drama, and dance together, showing 0.07 mean hours per week for adults in this category and 0.39 for tapes and records. The spread of compact discs, personal stereos, and MTV has complicated the issue since then. But the main point has likely remained the same.

20. Thomas S. Smith, "Late Nineteenth-Century Communications: Techniques and Machines," in Melvin Kranzberg and Carroll W. Pursell, Jr. (eds.), *Technology in Western Civilization* (New York: Oxford University Press, 1967), vol. 1, pp. 646–48.

21. Csikszentmihalyi and Rochberg-Halton, *The Meaning of Things*, pp. 71–72; see also pp. 243–44.

22. Ibid., p. 73.

23. Ibid., p. 115.

24. Ibid., p. 115.

25. Ibid., pp. 189–95.

26. Ibid., p. 193.

27. Ibid., pp. 195 and 243–44.

PART III

The Question of Heidegger

W INNER AND BORGMANN'S normative approach contrasts with Heidegger's ontological reformulation of the "question of technology." But although Heidegger's
contribution to the philosophy of technology is widely acknowledged, his work is controversial and difficult to understand. This section therefore endeavors to explain Heidegger's position, to document its growing influence in the technical fields themselves,
and to discuss its political implications.

According to Hubert L. Dreyfus, Heidegger was not concerned with the problem
of human control of technology that preoccupies the technology-as-ideology theorists.
In "Heidegger on Gaining a Free Relation to Technology," Dreyfus argues that the issue
is rather the understanding of being as mere raw material projected in the technological worldview of the modern West. Only by realizing that being itself commands this
dispensation can we begin to distance ourselves from that worldview. Ultimately, technology is not a political matter but a mode of givenness specific to the modern age.
This realization places us beyond a purely technological outlook on life and opens us
to a different relation to technology. What is needed is a new sense of reality that
privileges the non-technological facets of experience and community. In that context,
technology would continue to exist and do its work but would no longer form the
horizon of being and understanding.

In "Heidegger and the Design of Computer Systems," Terry Winograd discusses
Heidegger's influence on the world of computers. This is certainly unexpected, given
Heidegger's hostility toward cybernetics, but tends to confirm Dreyfus's analysis. Computer scientists and designers dissatisfied with the computational tradition have turned
to Heidegger not for a condemnation of technology but for new insights into how
better to fit it into human life. For the most part, they have been influenced by Heidegger's analysis of action in *Being and Time*. Winograd reviews several attempts to incorporate Heideggerian insights into artificial intelligence research, robotics, and interface design. He reviews Heidegger's significance for Scandinavian projects in
participatory design (also discussed by Winner), and he concludes with a discussion
of "ontological designing," or design as a method of communal self-understanding and
self-transformation. According to Winograd, in its implicit challenge to conventional
concepts in computer design, Heidegger's thought invites us to rethink our potential
for being human.

It is startling to realize that the Heidegger who inspired these two chapters was at
one time a Nazi and, perhaps more significantly, believed that national socialism of-

fered the only solution to the crisis of technological society. How deeply is his philosophy of technology implicated in this grotesque political and moral error? This is the question Tom Rockmore addresses in "Heidegger on Technology and Democracy." Rockmore concludes that understanding technology in terms of a transpersonal being rather than as a product of human purposes leads to the anti-humanist and undemocratic strain in Heidegger's thought.

6

Heidegger on Gaining
a Free Relation to Technology

Hubert L. Dreyfus

Introduction: What Heidegger Is Not Saying

IN *The Question Concerning Technology* Heidegger describes his aim:

> We shall be questioning concerning technology, and in so doing we should
> like to prepare a free relationship to it.

He wants to reveal the essence of technology in such a way that "in no way confines
us to a stultified compulsion to push on blindly with technology or, what comes to
the same thing, to rebel helplessly against it."[1] Indeed, he claims that "When we
once open ourselves expressly to the *essence* of technology, we find ourselves unexpectedly taken into a freeing claim."[2]

We will need to explain essence, opening, and freeing before we can understand
Heidegger here. But already Heidegger's project should alert us to the fact that he is
not announcing one more reactionary rebellion against technology, although many
respectable philosophers, including Jürgen Habermas, take him to be doing just that;
nor is he doing what progressive thinkers such as Habermas want him to do, proposing
a way to get technology under control so that it can serve our rationally chosen ends.

The difficulty in locating just where Heidegger stands on technology is no accident. Heidegger has not always been clear about what distinguishes his approach from
a romantic reaction to the domination of nature, and when he does finally arrive at a
clear formulation of his own original view, it is so radical that everyone is tempted to
translate it into conventional platitudes about the evils of technology. Thus Heidegger's
ontological concerns are mistakenly assimilated to humanistic worries about the devastation of nature.

Those who want to make Heidegger intelligible in terms of current anti-techno-
logical banalities can find support in his texts. During the war he attacks consumerism:

> The circularity of consumption for the sake of consumption is the sole pro-

cedure which distinctively characterizes the history of a world which has become an unworld.[3]

And as late as 1955 he holds that:

> The world now appears as an object open to the attacks of calculative
> thought. . . . Nature becomes a gigantic gasoline station, an energy source for
> modern technology and industry.[4]

In this address to the Schwartzwald peasants he also laments the appearance of
television antennae on their dwellings.

> Hourly and daily they are chained to radio and television. . . . All that with
> which modern techniques of communication stimulate, assail, and drive
> man—all that is already much closer to man today than his fields around his
> farmstead, closer than the sky over the earth, closer than the change from
> night to day, closer than the conventions and customs of his village, than the
> tradition of his native world.[5]

Such statements suggest that Heidegger is a Luddite who would like to return from
the exploitation of the earth, consumerism, and mass media to the world of the
pre-Socratic Greeks or the good old Schwartzwald peasants.

Heidegger's Ontological Approach to Technology

As his thinking develops, however, Heidegger does not deny these are serious
problems, but he comes to the surprising and provocative conclusion that focusing on
loss and destruction is still technological.

> All attempts to reckon existing reality . . . in terms of decline and loss, in terms
> of fate, catastrophe, and destruction, are merely technological behavior.[6]

Seeing our situation as posing a problem that must be solved by appropriate action
turns out to be technological too:

> [T]he instrumental conception of technology conditions every attempt to
> bring man into the right relation to technology. . . . The will to mastery be-
> comes all the more urgent the more technology threatens to slip from human
> control.[7]

Heidegger is clear this approach cannot work.

> No single man, no group of men, no commission of prominent statesmen,
> scientists, and technicians, no conference of leaders of commerce and indus-
> try, can brake or direct the progress of history in the atomic age.[8]

His view is both darker and more hopeful. He thinks there is a more dangerous situation facing modern man than the technological destruction of nature and civilization, yet a situation about which something *can* be done—at least indirectly. The threat is not a *problem* for which there can be a *solution* but an ontological *condition* from which we can be *saved*.

Heidegger's concern is the human distress caused by the *technological understanding of being*, rather than the destruction caused by specific technologies. Consequently, Heidegger distinguishes the current problems caused by technology—ecological destruction, nuclear danger, consumerism, etc.—from the devastation that would result if technology solved all our problems.

> What threatens man in his very nature is the . . . view that man, by the peaceful release, transformation, storage, and channeling of the energies of physical nature, could render the human condition . . . tolerable for everybody and happy in all respects.[9]

The "greatest danger" is that

> the approaching tide of technological revolution in the atomic age could so captivate, bewitch, dazzle, and beguile man that calculative thinking may someday come to be accepted and practiced *as the only* way of thinking.[10]

The danger, then, is not the destruction of nature or culture but a restriction in our way of thinking—a leveling of our understanding of being.

To evaluate this claim we must give content to what Heidegger means by an understanding of being. Let us take an example. Normally we deal with things, and even sometimes people, as resources to be used until no longer needed and then put aside. A styrofoam cup is a perfect example. When we want a hot or cold drink it does its job, and when we are through with it we throw it away. How different this understanding of an object is from what we can suppose to be the everyday Japanese understanding of a delicate teacup. The teacup does not preserve temperature as well as its plastic replacement, and it has to be washed and protected, but it is preserved from generation to generation for its beauty and its social meaning. It is hard to picture a tea ceremony around a styrofoam cup.

Note that the traditional Japanese understanding of what it is to be human (passive, contented, gentle, social, etc.) fits with their understanding of what it is to be a thing (delicate, beautiful, traditional, etc.). It would make no sense for us, who are active, independent, and aggressive—constantly striving to cultivate and satisfy our desires—to relate to things the way the Japanese do; or for the Japanese (before their understanding of being was interfered with by ours) to invent and prefer styrofoam teacups. In the same vein *we* tend to think of politics as the negotiation of individual desires while the Japanese seek consensus. In sum the social practices containing an understanding of what it is to be a human self, those containing an interpretation of what it is to be a thing, and those defining society fit together. They add up to an understanding of being.

The shared practices into which we are socialized, then, provide a background understanding of what counts as things, what counts as human beings, and ultimately what counts as real, on the basis of which we can direct our actions toward particular things and people. Thus the understanding of being creates what Heidegger calls a *clearing* in which things and people can show up for us. We do not produce the clearing. It produces us as the kind of human beings that we are. Heidegger describes the clearing as follows:

> [B]eyond what is, not away from it but before it, there is still something else that happens. In the midst of beings as a whole an open place occurs. There is a clearing, a lighting. . . . This open center is . . . not surrounded by what is; rather, the lighting center itself encircles all that is. . . . Only this clearing grants and guarantees to human beings a passage to those entities that we ourselves are not, and access to the being that we ourselves are.[11]

What, then, is the essence of technology, i.e., the technological understanding of being, i.e., the technological clearing, and how does opening ourselves to it give us a free relation to technological devices? To begin with, when we ask about the essence of technology we are able to see that Heidegger's question cannot be answered by defining technology. Technology is as old as civilization. Heidegger notes that it can be correctly defined as "a means and a human activity." He calls this "the instrumental and anthropological definition of technology."[12] But if we ask about the *essence* of technology (the technological understanding of being) we find that modern technology is "something completely different and . . . new."[13] Even different from using styrofoam cups to serve our desires. The essence of modern technology, Heidegger tells us, is to seek more and more flexibility and efficiency *simply for its own sake.* "[E]xpediting is always itself directed from the beginning . . . towards driving on to the maximum yield at the minimum expense."[14] That is, our only goal is optimization:

> Everywhere everything is ordered to stand by, to be immediately at hand, indeed to stand there just so that it may be on call for a further ordering. Whatever is ordered about in this way has its own standing. We call it standing-reserve. . . . [15]

No longer are we subjects turning nature into an object of exploitation:

> The subject-object relation thus reaches, for the first time, its pure "relational," i.e., ordering, character in which both the subject and the object are sucked up as standing-reserves.[16]

A modern airliner is not an object at all, but just a flexible and efficient cog in the transportation system.[17] (And passengers are presumably not subjects but merely resources to fill the planes.) Heidegger concludes: "Whatever stands by in the sense of standing-reserve no longer stands over against us as object."[18]

All ideas of serving God, society, our fellow men, or even our own calling disappear. Human beings, on this view, become a resource to be used, but more important to be enhanced—like any other.

> Man, who no longer conceals his character of being the most important raw material, is also drawn into this process.[19]

In the film *2001*, the robot HAL, when asked if he is happy on the mission, answers: "I'm using all my capacities to the maximum. What more could a rational entity desire?" This is a brilliant expression of what anyone would say who is in touch with our current understanding of being. We pursue the growth or development of our potential simply for its own sake—it is our only goal. The human potential movement perfectly expresses this technological understanding of being, as does the attempt to better organize the future use of our natural resources. We thus become part of a system which no one directs but which moves toward the total mobilization of all beings, even us. This is why Heidegger thinks the perfectly ordered society dedicated to the welfare of all is not the solution of our problems but the distressing culmination of the technological understanding of being.

What Then Can We Do?

But, of course, Heidegger uses and depends upon modern technological devices. He is no Luddite and he does not advocate a return to the pre-technological world.

> It would be foolish to attack technology blindly. It would be shortsighted to condemn it as the work of the devil. We depend on technical devices; they even challenge us to ever greater advances.[20]

Instead, Heidegger suggests that there is a way we can keep our technological devices and yet remain true to ourselves:

> We can affirm the unavoidable use of technical devices, and also deny them the right to dominate us, and so to warp, confuse, and lay waste our nature.[21]

To understand how this might be possible we need an illustration of Heidegger's important distinction between technology and the technological understanding of being. Again we can turn to Japan. In contemporary Japan a traditional, non-technological understanding of being still exists alongside the most advanced high-tech production and consumption. The TV set and the household gods share the same shelf—the styrofoam cup co-exists with the porcelain one. We can thus see that one can have technology without the technological understanding of being, so it becomes clear that the technological understanding of being can be dissociated from technological devices.

To make this dissociation, Heidegger holds, one must rethink the history of being in the West. Then one will see that although a technological understanding of being is our destiny, it is not our fate. That is, although our understanding of things and ourselves as resources to be ordered, enhanced, and used efficiently has been building up since Plato and dominates our practices, we are not stuck with it. It is not the way things have to be, but nothing more or less than our current cultural clearing.

Only those who think of Heidegger as opposing technology will be surprised at his next point. Once we see that technology is our latest understanding of being, we will be grateful for it. We did not make this clearing nor do we control it, but if it were not given to us to encounter things and ourselves as resources, nothing would show up *as* anything at all and no possibilities for action would make sense. And once we realize—in our practices, of course, not just in our heads—that we *receive* our technological understanding of being, we have stepped out of the technological understanding of being, for we then see that what is most important in our lives is not subject to efficient enhancement. This transformation in our sense of reality—this overcoming of calculative thinking—is precisely what Heideggerian thinking seeks to bring about. Heidegger seeks to show how we can recognize and thereby overcome our restricted, willful modern clearing precisely by recognizing our essential receptivity to it.

> [M]odern man must first and above all find his way back into the full breadth of the space proper to his essence. That essential space of man's essential being receives the dimension that unites it to something beyond itself . . . that is the way in which the safekeeping of being itself is given to belong to the essence of man as the one who is needed and used by being.[22]

But precisely how can we experience the technological understanding of being as a gift to which we are receptive? What is the phenomenon Heidegger is getting at? We can break out of the technological understanding of being whenever we find ourselves gathered by things rather than controlling them. When a thing like a celebratory meal, to take Heidegger's example, pulls our practices together and draws us in, we experience a focusing and a nearness that resists technological ordering. Even a technological object like a highway bridge, when experienced as a gathering and focusing of our practices, can help us resist the very technological ordering it furthers. Heidegger describes the bridge so as to bring out both its technological ordering function and its continuity with pre-technological things.

> The old stone bridge's humble brook crossing gives to the harvest wagon its passage from the fields into the village and carries the lumber cart from the field path to the road. The highway bridge is tied into the network of long-distance traffic, paced as calculated for maximum yield. Always and ever differ-

ently the bridge escorts the lingering and hastening ways of men to and fro.
. . . The bridge *gathers* to itself in *its own way* earth and sky, divinities and
mortals.[23]

Getting in sync with the highway bridge in its technological functioning can make
us sensitive to the technological understanding of being as the way our current
clearing works, so that we experience our role as receivers, and the importance of
receptivity, thereby freeing us from our compulsion to force all things into one efficient order.

This transformation in our understanding of being, unlike the slow process of
cleaning up the environment which is, of course, also necessary, would take place in
a sudden Gestalt switch.

The turning of the danger comes to pass suddenly. In this turning, the clearing
belonging to the essence of being suddenly clears itself and lights up.[24]

The danger, when grasped as the danger, becomes that which saves us. "The self-
same danger is, when it is *as* the danger, the saving power."[25]

This remarkable claim gives rise to two opposed ways of understanding Heidegger's response to technology. Both interpretations agree that once one recognizes the
technological understanding of being for what it is—a historical understanding—one
gains a free relation to it. We neither push forward technological efficiency as our only
goal nor always resist it. If we are free of the technological imperative we can, in each
case, discuss the pros and cons. As Heidegger puts it:

We let technical devices enter our daily life, and at the same time leave them
outside . . . as things which are nothing absolute but remain dependent upon
something higher [the clearing]. I would call this comportment toward technology which expresses "yes" and at the same time "no", by an old word,
releasement towards things.[26]

One way of understanding this proposal—represented here by Richard Rorty—
holds that once we get in the right relation to technology, viz. recognize it as a clearing,
it is revealed as just as good as any other clearing. Efficiency—getting the most out of
ourselves and everything else—is fine, so long as we do not think that efficiency for its
own sake is the *only* end for man, dictated by reality itself, to which all others must be
subordinated. Heidegger seems to support this acceptance of the technological understanding of being when he says:

That which shows itself and at the same time withdraws [i.e., the clearing] is
the essential trait of what we call the mystery. I call the comportment which
enables us to keep open to the meaning hidden in technology, *openness to
the mystery.* Releasement toward things and openness to the mystery belong

together. They grant us the possibility of dwelling in the world in a totally different way. They promise us a new ground and foundation upon which we can stand and endure in the world of technology without being imperiled by it.[27]

But acceptance of the mystery of the gift of understandings of being cannot be Heidegger's whole story, for he immediately adds:

> Releasement toward things and openness to the mystery give us a vision of a new rootedness which *someday* might even be fit to recapture the old and now rapidly disappearing rootedness in a changed form.[28]

We then look back at the preceding remark and realize *releasement* gives only a "possibility" and a "promise" of "dwelling in the world in a totally different way."

Mere openness to technology, it seems, leaves out much that Heidegger finds essential to human being: embeddedness in nature, nearness or localness, shared meaningful differences such as noble and ignoble, justice and injustice, salvation and damnation, mature and immature—to name those that have played important roles in our history. *Releasement*, while giving us a free relation to technology and protecting our nature from being distorted and distressed, cannot give us any of these.

For Heidegger, there are, then, two issues. One issue is clear:

> The issue is the saving of man's essential nature. Therefore, the issue is keeping meditative thinking alive.[29]

But that is not enough:

> If releasement toward things and openness to the mystery awaken within us, then we should arrive at a path that will lead to a new ground and foundation.[30]

Releasement, it turns out, is only a stage, a kind of holding pattern, awaiting a new understanding of being, which would give some content to our openness—what Heidegger calls a new rootedness. That is why each time Heidegger talks of *releasement* and the saving power of understanding technology as a gift he then goes on to talk of the divine.

> Only when man, in the disclosing coming-to-pass of the insight by which he himself is beheld . . . renounces human self-will . . . does he correspond in his essence to the claim of that insight. In thus corresponding man is gathered into his own, that he . . . may, as the mortal, look out toward the divine.[31]

The need for a new centeredness is reflected in Heidegger's famous remark in his last interview: "Only a god can save us now."[32] But what does this mean?

The Need for a God

Just preserving pre-technical practices, even if we could do it, would not give us what we need. The pre-technological practices no longer add up to a shared sense of reality and one cannot legislate a new understanding of being. For such practices to give meaning to our lives, and unite us in a community, they would have to be focused and held up to the practitioners. This function, which later Heidegger calls "truth setting itself to work," can be performed by what he calls a work of art. Heidegger takes the Greek temple as his illustration of an artwork working. The temple held up to the Greeks what was important, and so let there be heroes and slaves, victory and disgrace, disaster and blessing, and so on. People whose practices were manifested and focused by the temple had guidelines for leading good lives and avoiding bad ones. In the same way, the medieval cathedral made it possible to be a saint or a sinner by showing people the dimensions of salvation and damnation. In either case, one knew where one stood and what one had to do. Heidegger holds that "there must always be some being in the open [the clearing], something that is, in which the openness takes its stand and attains its constancy."[33]

We could call such special objects cultural paradigms. A cultural paradigm focuses and collects the scattered practices of a culture, unifies them into coherent possibilities for action, and holds them up to the people who can then act and relate to each other in terms of the shared exemplar.

When we see that for later Heidegger only those practices focused in a paradigm can establish what things can show up as and what it makes sense to do, we can see why he was pessimistic about salvaging aspects of the Enlightenment or reviving practices focused in the past. Heidegger would say that we should, indeed, try to preserve such practices, but they can save us only if they are radically transformed and integrated into a new understanding of reality. In addition we must learn to appreciate marginal practices—what Heidegger calls the saving power of insignificant things—practices such as friendship, backpacking into the wilderness, and drinking the local wine with friends. All these practices are marginal precisely because they are not efficient. They can, of course, be engaged in for the sake of health and greater efficiency. This expanding of technological efficiency is the greatest danger. But these saving practices could come together in a new cultural paradigm that held up to us a new way of doing things, thereby focusing a world in which formerly marginal practices were central and efficiency marginal. Such a new object or event that grounded a new understanding of reality Heidegger would call a new god. This is why he holds that "only another god can save us."[34]

Once one sees what is needed, one also sees that there is not much we can do to bring it about. A new sense of reality is not something that can be made the goal of a crash program like the moon flight—a paradigm of modern technological power. A hint of what such a new god might look like is offered by the music of the sixties. The Beatles, Bob Dylan, and other rock groups became for many the articulation of new understanding of what really mattered. This new understanding almost coalesced into a cultural paradigm in the Woodstock Music Festival, where people actually lived for a few days in an understanding of being in which mainline contemporary concern with rationality, sobriety, willful activity, and flexible, efficient control were made marginal and subservient to Greek virtues such as openness, enjoyment of nature, dancing, and Dionysian ecstasy along with a neglected Christian concern with peace, tolerance, and love of one's neighbor without desire and exclusivity. Technology was not smashed or denigrated but all the power of the electronic media was put at the service of the music which focused all the above concerns.

If enough people had found in Woodstock what they most cared about, and recognized that all the others shared this recognition, a new understanding of being might have coalesced and been stabilized. Of course, in retrospect we see that the concerns of the Woodstock generation were not broad and deep enough to resist technology and to sustain a culture. Still we are left with a hint of how a new cultural paradigm would work, and the realization that we must foster human receptivity and preserve the endangered species of pre-technological practices that remain in our culture, in the hope that one day they will be pulled together into a new paradigm, rich enough and resistant enough to give new meaningful directions to our lives.

To many, however, the idea of *a* god which will give us a unified but open community—one set of concerns which everyone shares if only as a focus of disagreement—sounds either unrealistic or dangerous. Heidegger would probably agree that its open democratic version looks increasingly unobtainable and that we have certainly seen that its closed totalitarian form can be disastrous. But Heidegger holds that given our historical essence—the kind of beings we have become during the history of our culture—such a community is necessary to us. This raises the question of whether our need for one community is, indeed, dictated by our historical essence, or whether the claim that we can't live without a centered and rooted culture is simply romantic nostalgia.

It is hard to know how one could decide such a question, but Heidegger has a message even for those who hold that we, in this pluralized modern world, should not expect and do not need one all-embracing community. Those who, from Dostoievsky, to the hippies, to Richard Rorty, think of communities as local enclaves in an otherwise impersonal society still owe us an account of what holds these local communities together. If Dostoievsky and Heidegger are right, each local community still needs its local god—its particular incarnation of what the community is up to. In that case we are again led to the view that releasement is not enough, and to the modified Heideggerian slogan that only some new *gods* can save us.

Notes and References

1. Martin Heidegger, "The Question Concerning Technology," *The Question Concerning Technology* (New York: Harper Colophon, 1977), pp. 25–26.

2. Ibid.

3. Heidegger, "Overcoming Metaphysics," *The End of Philosophy* (New York: Harper and Row, 1973), p. 107.

4. Heidegger, *Discourse on Thinking* (New York: Harper and Row, 1966), p. 50.

5. Ibid., p. 48.

6. Heidegger, "The Turning," *The Question Concerning Technology*, p. 48.

7. Heidegger, "The Question Concerning Technology," *The Question Concerning Technology*, p. 5.

8. Heidegger, *Discourse on Thinking*, p. 52.

9. Martin Heidegger, "What Are Poets For?" *Poetry, Language, Thought* (New York: Harper and Row, 1971), p. 116.

10. Heidegger, *Discourse on Thinking*, p. 56.

11. Heidegger, "The Origin of the Work of Art", *Poetry, Language, Thought*, p. 53.

12. Heidegger, "The Question Concerning Technology," p. 5.

13. Ibid.

14. Ibid., p. 15.

15. Ibid., p. 17.

16. Heidegger, "Science and Reflection," *The Question Concerning Technology*, p. 173.

17. Heidegger, "The Question Concerning Technology," p. 17.

18. Ibid.

19. Heidegger, "Overcoming Metaphysics," *The End of Philosophy*, p. 104.

20. Heidegger, *Discourse on Thinking*, p. 53.

21. Ibid., p. 54.

22. Heidegger, "The Turning," *The Question Concerning Technology*, p. 39.

23. Heidegger, *Poetry, Language, Thought*, pp. 152–53.

24. Ibid., p. 44.

25. Heidegger, "The Turning," *The Question Concerning Technology*, p. 39.

26. Heidegger, *Discourse on Thinking*, p. 54.

27. Ibid., p. 55.

28. Ibid. (My italics.)

29. Ibid., p. 56.

30. Ibid.

31. Heidegger, "The Turning," *The Question Concerning Technology*, p. 47.

32. "Nur noch ein Gott kann uns retten," *Der Spiegel*, May 31, 1976.

33. Heidegger, "The Origin of the Work of Art," *Poetry, Language, Thought*, p. 61.

34. This is an equally possible translation of the famous phrase from *Der Spiegel*.

7

Heidegger and the Design of Computer Systems

Terry Winograd

Introduction

A T FIRST GLANCE, computer systems might seem to be one of the least likely areas for applying Heidegger's insights. In an early article on phenomenology and artificial intelligence [9], Dreyfus describes the relation of computation to metaphysics as viewed by Heidegger:

> For Heidegger, technology, with its insistence on the "thorough-going calculability of objects," is the inevitable culmination of metaphysics, the concern with beings to the exclusion of Being. . . . Heidegger remarks that "the determination of language as information originally supplies the sufficient reason for the construction of thinking machines, and for the construction of large-scale computer installations," and that "information theory is, as pronouncement, already the arrangement whereby all objects are put in such form as to assure man's domination over the entire earth and even the planets." "Philosophy has come to an end in the present epoch. It has found its place in the scientific view . . . the fundamental characteristic of this scientific determination is that it is cybernetic, i.e., technological." (41)

However, beginning with that article and Dreyfus's other writings, there has been an ongoing confrontation between the computational and phenomenological traditions. Although most of the discourse has been a critique of what computers *can't do* (see [6]), a number of recent discussions have aimed at providing positive directions for design, with computer researchers explicitly seeking insights in Heidegger and pursuing those insights in their work.

It is a clear statement of the power of Heidegger's thought that even among a group of people from a very different tradition it has evoked stirrings—inklings of a new way of going about the business of technology. Before looking at detailed cases let us consider a few general points about the influence of Heidegger's ideas.

First, in describing the impact of Heidegger on computer scientists and designers, we are primarily referring to Division I of *Being and Time*. The larger breadth of issues that Heidegger addressed in his writing go far afield from the concerns of computer scientists in their technical role. Further, Heidegger's work has become known to the computing community almost exclusively through the interpretations of Hubert Dreyfus. His writings have been the most accessible, prominent, and visible and have been drawn on heavily in our own work as well as that of others described here.

Second, in noting the influence of Heidegger we find it entangled with many related influences. The researchers discussed below also cite a number of other sources, including Wittgenstein, the American pragmatists, Gibsonian psychology, Ethnomethodology, Soviet action theory, and so forth. Without trying to sort out all of the directions and differences, we can note that these represent loosely related traditions that are all being brought to bear on the question of how we can shift our understanding of computer design.

Finally, it is worth noting that the aspects of Heidegger's work that are being addressed are not the most radical ones—not the existential questions of Being—but rather the more practical everyday notions of readiness-to-hand, thrownness, breakdown, and the like. Although Heidegger's work raises fundamental questions about the nature of technology and the technological society, these have entered very little into the discourse of those who develop computer technology.

This chapter is not intended as a comprehensive survey of all of the ways in which Heidegger's work is influencing computer science and technology. It suggests possibilities by presenting examples in four areas of computer research and development, focusing on the work of a few individuals and groups. In each case, the work is explicit in its acknowledgment of Heidegger's influence and is representative of a new approach with respect to the previous background of related work. In each case it is useful to look at the tradition that this work was situated within and was reacting to in moving in a Heideggerian direction.

Artificial Intelligence

The first area of influence has been on work in artificial intelligence (that branch of computing that tries to duplicate human mental abilities). The background tradition in which most of this work has proceeded has been examined from a phenomenological perspective in a number of works [6, 8, 24]. The basic thrust is that AI derives almost totally from a mode of understanding language, thought, and action that is based on representation, formalism, and symbolic manipulations—a complex of related formulations which are best exemplified by the modern digital computer running programs organized around logical analysis.

Critics have questioned whether technology based on that approach can duplicate in any serious way human mental and linguistic capabilities. At the same time there are others, such as Preston, who turn this critique around, arguing that a new "Heideggerian AI" will succeed in matching the original dreams.

> I believe that ideas of great importance to AI and to the computational theory of mind in general are to be gleaned from Heidegger and from Dreyfus' critique of AI. Although this critique is often read as an argument against the mere possibility of artificial intelligence, it is better construed as an argument against a particular approach to the study of intelligent behavior, coupled with an attempt to sketch an alternative approach. ([19], 1)

Preston makes an appeal for serious investigation of what she calls "the Heideggerian alternative" and goes on to cite several examples in which she declares that "this alternative has been successfully put into practice."

The bulk of the work she describes is from the Artificial Intelligence Laboratory at the Massachusetts Institute of Technology. For those who have followed the history of artificial intelligence, it is ironic that this laboratory should become a cradle of "Heideggerian AI." It was at MIT that Dreyfus first formulated his critique, and for twenty years the intellectual atmosphere in the AI Lab was overtly hostile to recognizing the implications of what he said. Nevertheless, some of the work now being done at that laboratory seems to have been affected by Heidegger and Dreyfus.

Nonrepresentational Robotics

The first example is work on robotics being done by Rodney Brooks and his associates [4, 15]. The goals of this work are in the tradition of earlier robotics work, focused on the design of "intelligent" mechanisms.

> I wish to build completely autonomous mobile agents that coexist in the world with humans, and are seen by those humans as intelligent beings in their own right. ([4], 7)

Brooks's declared aim is to develop an "engineering methodology" for building these devices he calls "creatures," both for their practical uses and as a theory of intelligence:

> One goal is to gain knowledge about the nature of intelligence, and implementing theories through computation is deemed the most appropriate way because it ensures intellectual honesty. . . . We build our mobile robots in order to investigate intelligence and therefore learn about what it takes to build a system which we would consider clever. ([15], 2)

In its foundations, the development of computational models of intelligence is hardly Heideggerian. What, then, does Preston see as Heideggerian about this approach? Several points deserve note.

1. Decomposition by activity instead of by function.

> The fundamental decomposition of the intelligent system is not into independent information processing units which must interface with each other via representations. Instead the intelligent system is decomposed into independent and parallel activity producing [units] which all interface directly to the world through perception and action. . . . ([4], 1)

For Brooks an activity is "a pattern of interactions with the world" which is embodied in a subsystem of the device. Each activity or behavior-producing system individually connects sensing to action. Although the reification of this activity into a "unit" of the system is far from Heideggerian, the shift of emphasis from information to activity is in some sense parallel to the shift from focusing on the mental structure—the structure of intentionality—to the activity structure. Stretching Heideggerian terminology a bit, we might interpret it as moving from the present-at-hand to an analog of being-in-the-world—to a concern with the readiness of action.

2. Starting in the "real world" as opposed to abstraction.

Brooks places great emphasis on testing each component of his Creatures in the "real world." He argues that "At each step we should build complete intelligent systems that we let loose in the real world with real sensing and real action," hence his emphasis on mobile robots as distinct from other applications that have been favored in earlier work on AI.

In a section entitled "Abstraction as a dangerous weapon," he argues that previous AI systems went astray because of their use of micro-worlds that were abstracted away from the supposed domains of application. In this he clearly echoes Dreyfus's arguments about micro-worlds (see, for example, the preface to the second edition of *What Computers Can't Do* [6]).

It may be stretching things to say that this is a direct application of Heidegger, but there is a glimmering of the centrality of World and its irreducibility to formalizations.

3. Rejection of the primacy of representation.

The title of one of Brooks's key papers is "Intelligence without Representation" [4], and much of the discussion in it follows lines that have been argued in phenomenological critiques of AI.

> Artificial Intelligence has foundered on the issue of representation. When intelligence is approached in an incremental manner, with strict reliance on interfacing to the real world through perception and action, reliance on representation disappears. (1)
>
> . . . there need be no explicit representation of either the world or the intentions of the system to generate intelligent behaviors for a Creature.
>
> Even at a local level we do not have traditional AI representations. We never

use tokens which have any semantics that can be attached to them. . . . There are no variables that need instantiation in reasoning processes. There are no rules which need to be selected through pattern matching. There are no choices to be made. . . . (11)

We hypothesize that much of even human level activity is . . . a reflection of the world through very simple mechanisms without detailed representations. (12)

We see in this last quotation the tension between new directions (relinquishing the emphasis on representation) and the traditions in which they are embedded (the emphasis on simple mechanisms). We will see this further in discussing Agre's work below.

First, let us note Brooks's assessment of the Heideggerian influence on his approach. Brooks contrasts his approach to other related work in a series of short sections with titles such as "It Isn't Connectionism." One that stands out is entitled "It Isn't German Philosophy" (16).

In some circles much credence is given to Heidegger as one who understood the dynamics of existence. Our approach has certain similarities to work inspired by this German philosopher (e.g., Agre and Chapman 1986) but our work was not so inspired. It is based purely on engineering considerations.

In addition to reflecting the mood of disdain for philosophy that is in the background, this declaration illustrates the deep misinterpretation of cognition that permeates the whole culture of computer science and artificial intelligence.

The Analysis of Everyday Activity

The other work Brooks cites as being "inspired by this German philosopher" was also done at the MIT AI lab and is more explicit in acknowledging its debt to phenomenology. This work was done by the team of David Chapman and Philip Agre; in discussing it here we will draw primarily from Agre's 1988 dissertation, entitled "The Dynamic Structure of Everyday Life" [1].

While Brooks's work is framed as an argument against the AI tradition of representation, Chapman and Agre's is an argument against the tradition of planning and "problem-solving" [2]. Agre argues that "general-purpose Planning is a black hole attracting anyone who neglects to work out the dynamics of activity in the world of their interest" (29). He criticizes AI as adopting an:

. . . inappropriate metaphor system of inside and outside, boundary and contents. . . . The artificiality of mentalistic metaphors manifests itself in the need to "pass in" the entire situation to any process of making or evaluating Plans. [This leads to] an artificial effort to reconstruct the world within the agent's head, which must compensate for mentalism's artificially rigid boundary between inside and outside. . . . Ordinary activity is not amenable to such rigid *a priori* circumscriptions of relevance. (34)

Agre proposes replacing the planning and problem-solving approach with a "situated activity view."

[The thesis develops] a computational theory of routine activity and cognitive architecture . . . organized around a distinction between cognitive *machinery* and the *dynamics*—that is, recurring patterns of interaction—of activity in the world. (250)

For Agre, everyday activity is a product of routine practices, not organized according to predetermined plans. He states that "Everyday activity is fundamentally improvised; contingency is the central phenomenon. An agent conducts its everyday activity by continually redeciding what to do" (11).

As with Brooks, Agre's work is both critical of the cognitive tradition and still very much rooted within it. For someone trained in the traditions of AI, his claims reflect a radical reinterpretation in their emphasis on contingency and opportunistic action instead of rational goal-directed planning. At the same time they lie very much within the discourse of computational mechanisms, cognitive architectures, and decision making.

The greatest departure from the AI tradition in Agre's work is in his methodology. Rather than looking at carefully restricted laboratory tasks or artificial "microworlds," he devotes substantial effort to dissecting the detailed structure of activity in ordinary activities such as walking from his apartment to the subway. His focus, like Heidegger's, is on creating a clearing in which things can be seen in spite of their obviousness.

. . . it is difficult to appreciate the idea that everyday life is something that needs to be studied.

[The basis for understanding] is not theoretical abstraction but rather a spontaneous openness to the mind-blowing intricacy of humble phenomena. An awareness lost in projects and goals continually passes over these phenomena. To catch sight of them, one must, deep down, find one's own everyday life *interesting.* (15, emphasis in the original)

The opening here is very much pervaded by the Heideggerian concern with Dasein in its everyday existence. For Agre, this shift from simple robots to human everyday activities and routine skills leads to a fascination with detail, which is in the spirit of the sociological approach known as "ethnomethodology" (applied to computer systems in [21]). The "mind-blowing intricacy" becomes an object of wonder, standing in stark contrast to the brutal simplifications that are the stock-in-trade of standard AI methodology.

What is retained from conventional AI, though, is the adherence to the fundamental methodology of the computational paradigm: the search for explanations of human behavior in terms of mechanisms. Agre, like Brooks, argues for a new style of mechanism but is nevertheless grounded in the tradition of explaining behavior through the rule-governed operation of a mechanism—what he calls a "computational theory."

> The project of relating a theory of activity to a theory of machinery offers great promise and raises great difficulties. The promise is that a serious theory of everyday activity as a whole can provide a firm basis for the engineering analyses required by serious computational research, whether this research is intended to produce an explanatory theory of human beings or a justified design for an autonomous robot. (9)

The technical content of Agre's dissertation abandons the realm of everyday activity to describe instead the workings of a series of programs that operate in the highly abstract "blocks worlds" typical to AI, including a program named Pengi [2], designed and engineered primarily by David Chapman, that plays a video game. In contrast to Brooks's emphasis on the "real world," Chapman's program does not deal with the physical activities of actual video games, but operates in a computational space of representations of game boards and actions.

Although Agre asserts that "Pengi is a big improvement on the blocks world" (196), he acknowledges that it "bears only a glancing resemblance to any activity in the physical world" (173). There is a huge gulf between his open-textured accounts of everyday life and Pengi's fully defined, well-constrained list of objects, properties, and actions. The motivation for accepting this kind of artificial domain is clear—it was necessary for the sake of building a working system.

Agre's critique is not of the overall goal of computational modeling, but rather of the specific approaches that have been taken, in particular the proliferation of complex cognitive mechanisms constructed on the model of conscious symbolic reasoning:

> I will argue that the principal task of artificial intelligence research is not the proliferation of complex forms of machinery but rather the elucidation of the dynamic structure of complex forms of activity. Having developed an understanding of dynamics, one should seek the simplest machinery that is consistent with the forms of activity one wishes to produce or explain. (12)

Motivated in part by Heidegger's rejection of present-at-hand reflection as the basis of action, he proposes a simple kind of "wiring" that does not depend on a theory of representation. He proposes a number of technical devices that are suggested, at least metaphorically, by some of Heidegger's arguments (which he cites). For example, he describes a mechanism for implementation he calls "deictic representation" as having been influenced by Heidegger's "analysis of *worldhood*, in particular the notions of *significance and assignment.*"

A system employing deictic representation can be built with simple machinery [dependency networks] because it can achieve abstraction without recourse to pattern matching and variable binding.

In summary: properly implemented, dependency networks are blindingly fast, massively parallel, and easy to construct. . . . The big job is to demonstrate that these simple accounts suffice. (77)

Usability Engineering

Turning away from the focus on how machines can be made intelligent, we can ask something else of the technology: How do people find computer systems usable? In work on human-computer interaction, the concern is not whether machines are like people in their thought and actions, but whether they constitute tools that can be used effectively.

In this general area, we find a large gap between theory and practice. On the one hand, there is a highly developed and successful body of practice in making usable systems. Practical know-how is instantiated by "virtuoso designers" who come up with both incremental improvements and substantial leaps, such as the graphic-oriented interfaces developed by Alan Kay and his colleagues at Xerox PARC and made popular by the Apple Macintosh. In identifying and evaluating such advances people look not to theory but to the marketplace—a design is successful when it is widely used and copied by other manufacturers.

The theoretical discourse about usability is dominated by a background based on a narrow kind of cognitivism. Users and their activities are characterized in terms of tasks, information, and mental operations. For example, in one of the standard texts on user interface design [20], Shneiderman offers an analysis of the four stages of an interaction:

1. Forming an intention: internal mental characterization of a goal.
2. Selecting an action: review possible actions and select most appropriate.
3. Executing the action: carry out the action using the computer.
4. Evaluating the outcome: check the results of executing the action.

This is offered as a model of what happens when someone hits a key on the computer. It is completely foreign to a Heideggerian approach in its manifestly present-at-hand articulation of mental acts such as "identify," "select," "choose" and "review." Working from this orientation, a variety of formal models and experimental methods have been developed.

Typically, experimenters put subjects in front of a computer screen and ask them to perform well-circumscribed tasks, such as making a certain set of editing changes to a document. The experiments measure variables such as how fast various tasks are performed with different kinds of command structures, pointing devices, and the like.

In reaction against this tradition, some researchers have advocated more phenomenological approaches to "usability engineering." Whiteside, for example, advocates what he calls "contextual research" [22, 23]. He and his group at the Digital Equipment Corporation consult on product designs within the company to see what will make them usable (ready-to-hand) given the overall structure of the product goals and user community. For example, they were successful in greatly increasing the perceived usability of a text editor developed for use by programmers [23]. They work within a domain of distinctions that is explicitly phenomenological, examining the background of practices and meanings that users bring to the technology they confront.

> The contextualist is precisely interested in what is unique about each situation. . . .
> Any analysis takes place within a context. . . . Heidegger, on whom our work . . . is explicitly based, uses contextualist distinctions. For him the most important mode of experience is the practical, ready-to-hand, ongoing involvement in a world of human concern. The detached analysis of the world in terms of abstract properties, which is a principal method of formism and mechanism, is NOT primary in Heidegger's scheme. Instead, the starting point is ongoing experience, in the moment experienced, in the particular context experienced, prior to explanation or analysis in terms of properties. ([22], 371)

For the contextualist, usability has its basis in the phenomenology of the user's experience of usability. The focus is on looking at people not in laboratories but in the real world, and studying them not by clocking their reactions but by talking with them about what they are doing.

> The purpose of contextual research methods is to uncover the user's experience of usability: that is, to identify dimensions of usability important to the user. Only users know their own experience of the product, how they are using tools to do their work, and their perceived requirements of their work. Because experience is internal and unobservable, the usability engineer cannot know the user's experience of a system through observation alone. Throughout a contextual research interview, usability engineers will generate interpretations of the user's behavior. But the usability engineer cannot know if conclusions are warranted unless he or she shares these interpretations with the user. . . . Through mutual interpretation the usability engineer and user come to a shared understanding of the participant's experience of the product. . . . The users become co-researchers. ([23], 25)

The designer and user enter into a hermeneutic dialogue, in which they are merging distinct horizons and generating a new understanding. Whiteside's discussion is pervaded by the use of Heideggerian terms like "concealment" and the focus on what might be called "concern." It also draws on his analysis of the way that objects and properties come into existence through language in the face of breakdown. In a chapter for the *Handbook of Human-Computer Interaction* [23], Whiteside and his colleagues pose a series of rhetorical questions about the methodology for creating usable systems.

> Could we gain a better understanding of usability if we collected our data within the context of the users' real work? . . . human action as observed derives its meaning from the context in which it occurs . . . much usability research takes place within the laboratory context. How might altering the context alter the meaning and the significance of the behavior observed? . . . In the laboratory, subjects perform tasks prescribed by the experimenter. In the workplace, people perform tasks important to their careers and livelihood. What shows up and what is concealed against these different backgrounds? (19)

One example is given of a project in which a notion of "control" came to be a primary feature of usability, after it emerged from discussions with users of a system they were developing.

> Once this construct ["control"] was identified, the usability engineer was able to present it to other users to obtain their response. When these users began to pick up the word, use the word spontaneously, and explain their experience in terms of that concept, we knew that "control" was a key usability concept. . . . The emergence of usability concepts has direct implications for design. Once such articulated concepts are available, they may be given operationally defined criteria in order to become part of usability specifications and evaluation processes. More importantly usability concepts begin to form a way of thinking about and a language for describing usability. (28)

> Through developing measurable usability specifications, negotiating a shared vision for usability, and making the results known to all involved in the project, we have a basis for conducting the business of engineering. The usability specifications . . . constitute an objective, modifiable, and public vision of what the group is trying to achieve. (5)

In this chapter, we note that the hermeneutic insight that the concepts emerge in the interaction rather than existing in the machine or the mind of the user is coupled with the quest for "objective" and "operationally defined criteria" very much within the tradition of engineering. This work combines the hermeneutic approach with a more traditional process of quantified analysis, trying as with the artificial intelligence approaches described above to apply Heideggerian analysis within the overall accepted framework.

System Development Methodologies

The emphasis on a hermeneutic interaction between the designer and user has been a key focus of a body of work on systems development methodologies that has grown up over the past few years in Scandinavia [10, 3].

The conventional practice for computer system development is called the "waterfall model." It consists of a cascaded sequence of steps in which designers first ask the purchasers of the system what it is they want (a "requirements document"), then generate a formal specification of the functioning of the system, then ship that specification off to someone who will write the program. The finished program is tested to some extent against the specification, then is shipped back, installed, and employed. This methodology concentrates on getting the right formal description of what the system will do, often in a form required by various software contracting protocols and agreements, particularly for military purchasers.

Against that background, the advocates of an alternative approach, which they call "participatory design," have argued that design is a hermeneutic activity, in which the nature of the finished system emerges from a process of developing shared interpretations among the relevant parties, including designers, managers, and the workers who will be in direct contact with the resulting devices. In their analysis they make explicit reference to the philosophy of Heidegger. For example a recent conference was entitled "Software Development and Reality Construction: An Invited Working Conference Focusing on Epistemological Foundations for Development and Use of Computer Based Systems" [5]. It included, among other sessions, one called the "Heidegger working group" which included a presentation by Goguen on "Heidegger and Formal Semantics" [16].

> The failure of the waterfall model to even acknowledge the possibility of error is a shocking example of arrogant teleological thinking run wild; even some crude form of feedback control would be an improvement. . . . But we can, and should, go further than merely recognizing the inevitability of error—we can learn to experience our errors as a path that leads us to deeper understandings and to better relationships.

Goguen quotes from Heidegger's *Introduction to Metaphysics* [18]:

> The area, as it were, which opens in the interwovenness of being, unconcealment, and appearance—this area I understand as ERROR. Appearance, deception, illusion, error stand in definite essential and dynamic relations which have long been misinterpreted by psychology and epistemology and

which consequently, in our daily lives, we have well nigh ceased to experience and recognize as powers. (109)

Participatory design was the focus of several major projects in Denmark, Norway, and Sweden, including one called UTOPIA carried out at the Swedish Center for Working Life in conjunction with the unions representing the typesetting and graphic layout workers in the Swedish newspaper publishing industry. A number of researchers were involved, and I will primarily refer to the description and accompanying theoretical justification by Ehn [10].

The researchers worked with the union in developing new tools for quality graphics production, working explicitly from a theory based on aspects of Heidegger's philosophy, along with Wittgenstein, Marx, and others. Their focus was on grappling with questions of background and readiness-to-hand in design, as pointed to by Heidegger.

In the following inquiry into design and use of computer artifacts I suggest as an alternative point of departure an ontological and epistemological position that focuses on human *practice* to replace the dualist mirror-image theory of reality. . . . Practice is ontological. It is the human form of life. To be-in-the-world is more fundamental than subject-object relations. ([10], 60)

. . . a main concern in design of computer artifacts must be to design for future situations of "thrownness of unhampered" *use*. We should then aim at designing computer artifacts that are ready-to-hand for the users in their ordinary use situations. (66)

Ehn poses two basic dilemmas to be solved: the "gulf between the designer and user" and the tension between what he calls "tradition and transcendence."

First, in looking at the interaction between designer and user he addresses the same problem that drives the contextual research described in the previous section. Participatory design calls for a thoroughgoing openness, in which the design process does not even begin with a specification of a set of functions for the computer devices, but rather starts from the overall work situation, asking what it makes sense to computerize. To design well it is necessary to work with the interconnection between the tools and the work as a set of practices and know-how. As the UTOPIA researchers noted, the background in which a computer expert interprets the work of producing a newspaper is different from that of the graphics worker who has a horizon developed over years of observation and practice.

How should [the designer] be able to design the computer artifacts ready-to-hand for the users, when he does not understand the profession? Here he may create most undesirable breakdowns. (79)

. . . Principally, there seems to be two ways out of the dilemma. . . . by acquiring proficiency or expertness in the domain we are designing for we may be able to understand how to design computer artifacts as skill enhancing tools for the experts, without having to make the skill explicit, formalized and routinized. The designer becomes involved in thrown ready-to-hand activity *both as design and as use.*

The complementary way to surpass breakdown by formal descriptions may be to create conditions for skilled users to utilize their professional skills in the design process. The proficiency and expertness we are looking for in design may be mastery of methods that let the user participate in design as if it was his or her thrown ready-to-hand use activity. ([10], 75)

The second dilemma Ehn describes is between tradition and transcendence:

. . . we also must be able to deal with the following contradiction:
On the one hand, to design so as not to break down or make obsolete the understanding and readiness-to-hand the users have acquired in the use of the already existing artifacts. The new artifacts should be ready-to-hand in an already existing practice.
On the other hand, to break down the understanding of the already existing situation and make it present-at-hand, is to make reflection about it possible, and hence to create openings for a new understanding and alternative designs. ([10], 77)

His resolution is to apply methods in which design interventions grow out of experience. He advocates that designers share the users' practice for a long time and use what he calls "design by doing" methods, such as building partial prototypes, doing field tests with experimental systems, and having users develop mock-ups. Rather than trying to initially set out what is to be done, as in the waterfall model, the aim is to give people the opportunity to go through a process of trial and error in which what emerges can be thought of as the specification. The focus is on iterative and participatory design methodologies centered not on theory, but on the use of "design artifacts" such as prototypes and mock-ups.

Design artifacts, linguistic or not, may in a Wittgensteinian approach certainly be used to create breakdowns, but they must make sense in the users' ordinary language-games. If the design artifacts are good, it is because they help users and designers to see new aspects of an already well-known practice, not because they convey a theoretical interpretation. . . . (113)

Projects such as UTOPIA have developed a methodology for the social process of design, in which the designers become workers and the workers become designers as part of a long-term process out of which grow the designs. The focus has been not on theoretical analysis of the design process but on the practical political and social concerns that come up in creating appropriate groups, developing trust, working together, building suitable design artifacts, and declaring the outcomes.

Ontological Design

The last example takes a deeper direction from Heidegger, seeing the design of technology as a generative act that touches on the nature of being human. It has been developed in a series of educational and computer designs by Flores and his co-workers [11, 12, 13, 14, 24].

Like Whiteside and Ehn, Flores begins with the question of design. But he rejects their claim that the primary focus should be on the social methodology—on the situated participation of users in the design process:

> We challenge the received wisdom and cherished notions about design and communication. We wish to refer in particular to two erroneous notions.
> The first is the common belief that design should be dominated by the desires of the *user*. Prejudices derived from this belief are (a) that the best way to discover what the user wants is by asking questions by means of interviews and by observing and studying the user, and (b) that the criteria for design will evolve inductively after a trial and error process. We believe that this approach is wrong. The structure of the interactions is not chosen or agreed upon by the users, but rather must reflect the structure of the deeds performed. The deeds are not independent of language and vice versa. The systems designer needs to understand this structure *before* starting to study and design; questions are important, but only if the researcher knows what to ask. ([14], 96)

Flores rejects the overall tradition of design that begins with the question of equipment as a carrier of functions as seen by the users. He begins instead with the question of what it is to be human and, like Heidegger, pursues it as the means of creating a clearing in which understanding and new design can emerge. His concern is with unconcealing the fundamental ontology that underlies the use of computers, in particular the use of computers in work.

> The principal task is that of finding a proper way to analyze and design communication processes in an office. The design of specific equipment is a derivative part of this fundamental design task. ([11], 66)

In approaching the design of computer systems for use in work situations, Flores seeks the ontological basis of work by people together:

> In our investigation the crucial question seems to be "What is work?" One obvious answer at this point seems to be "communication", but the question then becomes a new question, "What is communication?" ([11], 66)

Flores argues for the need to move away from the rationalistic tradition that has dominated the discourse about computing and about language and action in general, and to develop new foundations. In this effort he begins with Heidegger's analysis, and then moves in a new direction.

> Coping with equipment and things is called by Heidegger *concern*. Dealing with people is called *solicitude*. The whole structure of coping with the world is called *care*. Heidegger describes the structure of *concern* that is manifested in the structure of referentiality, which is in turn revealed by the prepositions of language. However, he lacks a rich framework to speak about *solicitude* with the same precision; in order to do this it is necessary to have an interpretation and a framework for referring to social relationships, such as the one provided by the speech-acts theory as developed by Austin and Searle. One of our purposes in this and other works has been to pursue further the study of the structure of social connectivity, with the aim of discovering the minimal communicative tools that one must have in every moment of interaction. We have produced our synthesis between these two different and apparently conflicting philosophical traditions with this design aim in mind. ([14], 102)

Instead of taking Dasein as the kind of Being that is concerned with the question of Being, Flores addresses human Being as the kind of Being that exists in a world that it creates through commitment expressed in acts of language.

> People in an office participate in the creation and maintenance of a process of communication. At the core of this process is the performance of linguistic acts that bring forth different kinds of commitments. ([14], 105)

> We may say in a simple way that managers create commitments in their world, take care of commitments, and initiate new commitments within the organization. But, in our statement, *world* is also what is brought forth through language as a commitment established by an utterance. ([14], F109)

> In the framework of our theory, organizations are institutional settings that orient the structure of commitment. ([14], F96)

This foundation was the basis for the design of a computer tool for communication called The Coordinator, which is currently employed by more than 100,000 users. The Coordinator is a system for managing action in time, grounded in a theory of linguistic commitment and completion of conversations. It opens up a space of communicative actions, within the context of a computer network, which reflect the fundamental ontology of language and conversation.

> . . . while The Coordinator exemplifies a new design and a new theory of action and management as a basis for design, the distinctions of linguistic acts and completion of action are not distinctions of new entities or new proposals for doing something. What we are doing in our theory is reconstructing constitutive distinctions of human social action. These are distinctions for generating any socially coordinated actions: bringing—in a request—a future

action and its conditions of fulfillment into a publicly shared world; and producing—in a promise—a commitment to complete the action. These distinctions are simple, universal, and generative of the complex organizational and management phenomena with which we need to deal. ([13], 163)

In addition to starting design with an analysis of the ontology, Flores points out that design interventions create new ways of Being.

The most important designing is *ontological*. It constitutes an intervention in the background of our heritage, growing out of our already-existent ways of being in the world, and deeply affecting the kinds of beings that we are. In creating new artifacts, equipment, buildings, and organizational structures, it attempts to specify in advance how and where breakdowns will show up in our everyday practices and in the tools we use, opening up new spaces in which we can work and play. Ontologically oriented design is therefore necessarily both reflective and political, looking backwards to the tradition that has formed us but also forwards to as-yet-uncreated transformations of our lives together. Through the emergence of new tools, we come to a changing awareness of human nature and human action, which in turn leads to new technological development. The designing process is part of this "dance" in which our structure of possibilities is generated. ([24], 163)

Flores argues that the theory is the starting point for driving the process of design. He recognizes that the role of theory in creating a new clearing is more powerful than the unexamined merging of horizons that comes from a focus on methods for participation and discussion. The ontological designer needs to begin with an orientation to the kinds of questioning that will unconceal the ontology of the work, rather than posing technical questions about what the equipment can do. In this he has an answer to Ehn's dilemma of tradition and transcendence.

The practical consequence of our investigations for the design of organizations is that it is possible to have a theory which precedes the empirical study of organizations. ([11], 97)

Our principal theoretical claim is that human beings are fundamentally linguistic beings: action happens in language, in a world constituted through language. What is special about human beings is that they produce, in language, common distinctions for taking action together.

In turning our attention to this ontology, we are not designing something new for human beings to do. People already produce a world together in language and they already coordinate their actions in that world. A fundamental condition of human action is the ability to affect and anticipate the behavior of others through language. Design can improve the capacity of people to act by producing a reorganization of practices in coherence with this essential, ineliminable nature of human interaction and cooperation. ([13], 156)

The role of the designer includes the job of making the theory available to the people who will be affected, not just as a way of shaping the computer artifacts, but as a tool for redesigning their work and their lives.

> We are not asserting that people are aware of what they are doing; they are simply working and speaking, more or less blind to the pervasiveness of a commitment's essential dimensions. It is because of this peculiar fact that one of our recommendations for organizational design takes the following form:
> The process of communication should be designed to bring with it a major awareness about the occurrence of *commitments*. Every member's knowledge about his participation in the network of commitment must be reinforced and developed. ([14], 106)

Unlike the other researchers cited above, Flores is not a computer scientist and he approaches Heidegger from a different horizon. In his work he is not asking "What can computers do?" or "how can they do it?" but "What do people do?" "How can we understand computers as more than tools but as carriers of new ways of being?"

> In ontological designing, we are doing more than asking what can be built. We are engaging in a philosophical discourse about the self—about what we can do and what we can be. Tools are fundamental to action, and through our actions we generate the world. The transformation we are concerned with is not a technical one, but a continuing evolution of how we understand our surroundings and ourselves—of how we continue becoming the beings that we are. ([24], 179)

Conclusion

These four examples of Heidegger's influence on computer system design represent a beginning. As we continue to move beyond the early spectacular successes of computing technology, the inevitable "slowdown" will lead computer scientists and practitioners to take a more thoughtful look at the foundations underlying their successes and their failures. They will move away from the unexamined rationalistic discourse that has been taken for granted, and Heidegger's work will play a growing role in developing a new clearing for future design.

On the other hand, it would hardly be accurate to imply that there has been a wave of Heideggerian influence in computer science, or that the followers of Descartes, Boole, and Turing are in danger of seeing the foundations they laid abandoned. The work described here is relatively unique and is not yet in the center of the discourse about computers and their use.

Looking ahead, we can anticipate three styles in which Heidegger's philosophy will have an impact on the design of computer systems.

First, it has already been influential in offering a critique of current approaches and in grounding an assessment of the the limitations of the systems that are being proposed and built. Hubert Dreyfus has been a highly visible commentator on work in artificial intelligence [6], and his recent work with Stuart Dreyfus has addressed the current applied use of AI techniques in "expert systems" [8]. Many of his arguments that were initially derided by the computer technologists as uninformed speculations are now a serious part of the thinking about these applications.

The second direction is in suggesting methods for design that are more hermeneutic in their nature and more oriented to producing readiness-to-hand. In this area, much of the work we might describe as revealing a Heideggerian influence is in line with trends that were already present in the practices (though not the academic theory) of the field. The response of many people engaged in practical design to some of the arguments made by the various researchers described above would be "That's not Heidegger, it's just common sense." This is not a criticism, but a validation. Heidegger's analysis works because it reveals the facticity of the world we already inhabit. At the same time, there is a need to make it explicit. The "common sense" it reveals has long been concealed by the descendants of the Cartesian tradition, and needs to be brought back to light.

The third and ultimately most significant direction lies in the shift from ontic to ontological design—moving beyond the design of equipment to the design of Being through our activities of bringing forth. Ontological design is a harder task and one that must address deep forms of concealment and resistance.

> Dasein's kind of being thus demands that any ontological interpretation which sets itself the goal of exhibiting the phenomena in their primordiality, should capture the being of the entity, in spite of this entity's own tendency to cover things up. Existential analysis, therefore, constantly has the character of doing violence whether to the claims of the everyday interpretation, or to its complacency and its tranquilized obliviousness. ([17], 359)

Organizations (including organizations that use computers) are full of tranquilized obliviousness, and awareness and change are difficult tasks, not just the job of developing a new machine or formal methodology.

The challenge in applying Heidegger's philosophy to computation is not just as a way of building more effective computer systems, but in understanding computers as part of the network of equipment within which we encounter our Being. Heidegger offers us much more than the opportunity to improve our programming techniques. He challenges us to reexamine our being, not as detached observers, but as we create ourselves in our everyday lives and work. This challenge does not leave out our work as computer system builders and designers, and we have the potential to use our work in the difficult but deeply significant task of participating in Dasein's ongoing design of the potential for being human.

References

[1] Agre, Philip, "The Dynamic Structure of Everyday Life," MIT AI Lab Technical report 1085, 1988 (dissertation).

[2] Agre, Philip, and David Chapman, "Pengi: An Implementation of a Theory of Activity," AAAI-87.

[3] Bjerknes, Gro, Pelle Ehn, and Morten Kyng, *Computers and Democracy*, Aldershot: Avebury, 1987.

[4] Brooks, Rodney, "Intelligence without Representation," MIT AI Lab, manuscript April 22, 1987.

[5] "Software Development and Reality Construction: An Invited Working Conference Focusing on Epistemological Foundations for Development and Use of Computer Based Systems," Schloss Eringerfeld, Germany, Sept. 26–30, 1988.

[6] Dreyfus, H. L., *What Computers Can't Do: A Critique of Artificial Reason*, 2nd edition, New York: Harper and Row, 1979.

[7] Dreyfus, Hubert, and Stuart Dreyfus, "Making a Mind vs. Modeling the Brain," *Daedalus* 117:1 (Winter 1988): 15–44.

[8] Dreyfus, Hubert L., and Stuart E. Dreyfus, *Mind over Machine*, New York: Macmillan/The Free Press, 1985.

[9] "Phenomenology and Artificial Intelligence," in J. M. Edie (ed.), *Phenomenology in America*, Quadrangle Books, 1967, pp. 31–47.

[10] Ehn, Pelle, *Work-Oriented Design of Computer Artifacts*, Stockholm: Arbetslivscentrum, 1988.

[11] Flores, C. Fernando, "Management and Communication in the Office of the Future," doctoral dissertation, University of California at Berkeley, 1981.

[12] Flores, Fernando, and Michael Graves, "Domains of Permanent Human Concerns" and "Education," unpublished reports, Emeryville, CA: Logonet Inc., 1986.

[13] Flores, Fernando, Michael Graves, Bradley Hartfield, and Terry Winograd, "Computer Systems and the Design of Organizational Interaction," *ACM Transactions on Office Information Systems* 6:2 (April 1988), pp. 153–72.

[14] Flores, C. Fernando, and Juan Ludlow, "Doing and Speaking in the Office," in G. Fick and R. Sprague (eds.), *DSS: Issues and Challenges*, London: Pergamon Press, 1981.

[15] Flynn, Anita, and Rodney Brooks, "MIT Mobile Robots—What's Next?," working paper 302, MIT Artificial Intelligence Laboratory, November, 1987.

[16] Goguen, Joseph, "Heidegger and Formal Semantics," position paper for the *Working Conference on Software Development and Reality Construction*, 1988.

[17] Heidegger, Martin, *Being and Time*, translated by John Macquarrie and Edward Robinson, New York: Harper and Row, 1962.

[18] Heidegger, Martin, *Introduction to Metaphysics*, translated by Ralph Manheim, New Haven: Yale University Press, 1959.

[19] Preston, Beth, "Heidegger and Artificial Intelligence," manuscript, presented at Western Division APA Conference, 1989.

[20] Shneiderman, Ben, *Designing the User Interface: Strategies for Effective Human-Computer Interaction*, Reading, MA: Addison-Wesley, 1987.

[21] Suchman, Lucy, *Plans and Situated Actions: The Problem of Human-Machine Communication*, Cambridge: Cambridge University Press, 1987.

[22] Whiteside, John, and Dennis Wixon, "Contextualism as a World View for the Reformation of Meetings," *Proceedings of the Second Conference on Computer-Supported Cooperative Work*, Association for Computing Machinery, 1988, 369–76.

[23] Whiteside, John, John Bennett, and Karen Holtzblatt, "Usability Engineering: Our Experience and Evolution," in Martin Helander (ed.), *Handbook of Human-Computer Interaction*, North Holland, 1989.

[24] Winograd, Terry, and Fernando Flores, *Understanding Computers and Cognition: A New Foundation for Design*, Norwood, NJ: Ablex, 1986. Paperback issued by Addison-Wesley, 1987.

8

Heidegger on Technology and Democracy

Tom Rockmore

ONE OF THE most interesting and troubling facts about Martin Heidegger, a realization brought forcefully to our attention by the publication of Victor Farias's
study of Heidegger and nazism, is the fact that, after all the qualifications and interpretation, this important German thinker was deeply and irrevocably committed to one
of the most anti-democratic movements of this or any other historical moment.[1]

The problems Farias expounds had been identified previously by others, including
the gifted but dogmatic Marxist Georg Lukács. One is tempted to dismiss Lukács's
opinion because of his dogmatic attitude, but this would be a mistake. On examination, we see that his early, in fact nearly prescient, reading of the problem of Heidegger's relation to fascism looks better now than it did when Lukács wrote.[2] Unlike some
others, who at this late date often still hesitate to draw clear conclusions about Heidegger, Lukács was never constrained by such intellectual niceties.[3] One need not accept
his harsh verdict on Heidegger as the Ash Wednesday of subjective idealism.[4] Lukács
is also incorrect in labeling Heidegger as merely pro-fascist in view of the latter's public
and apparently sincere commitment to the essence of national socialism, as witness
his famous reaffirmation of the significance of the movement as late as 1953.[5] This
situation invites sober reflection on the extent to which claims for the social relevance
of philosophic reason, claims that run throughout the philosophic tradition from Socrates to the present, are realized in the case of this brilliant but sociopathic thinker. In
particular, we need to inquire if Heidegger's somber pessimism is justified with respect
to the phenomenon of technology.[6] And we need to join Lukács in raising an equally
important issue, namely whether Heidegger, under the cover of the question of the
meaning of Being, has not transformed, in mythological fashion, the human condition
in advanced capitalism into a universal and general human condition.[7]

A caveat is in order. In the wake of our knowledge of Heidegger's relation to national socialism, it is not useful, nor is it my intention, to besmirch his reputation in any
way. In any case, no critic of his position could do so more or more effectively than he
himself has already done. Rather, we need now to rethink our attitude toward his

thought without any preconditions in order to determine if, as some believe, the thinker and his thought are independent entities, or if, as others hold, the two are inextricably intertwined. Obviously this is a complicated task, one that will require many years of careful discussion among philosophers. It would be as irresponsible as Heidegger himself appears to have been for us to prejudge the case without providing an adequate hearing. But it would be equally irresponsible to fail to raise questions of this kind. Indeed, Heidegger seems to invite this type of examination of his thought on specific grounds, including the general insistence on *Dasein* as inevitably situated and the specific claim, in the course of the *Spiegel-Gespräch,* that his own thought must be judged in terms of his political engagements.

The aim of this chapter is to begin to undertake this task as concerns Heidegger's view of technology. It is this same view of technology which has recently been described by a revisionary critic as the most powerful part of Heidegger's corpus where everything comes together.[8] We need to decide whether Heidegger's reading of technology can stand as a permanent contribution by an important thinker. We need to ask ourselves whether this reading is adequate or fundamentally flawed.

Four points will help to situate Heidegger's notion of technology within its conceptual and historical context. First, the discussion of technology occurs in Heidegger's thought after the end of the war. This discussion, which follows Germany's defeat in the Second World War, exhibits a deep pessimism, particularly in terms of the relation of human beings to technology. But we will need to determine if the pessimism which is almost a constitutive part of Heidegger's thought in both its first and second periods is based on anything more than his subjective reflection of the existential situation he described in *Being and Time.* Thus, we will need to evaluate the reasons Heidegger advances to justify his pessimistic perspective.

Second, explicit attention to the concept of technology engages Heidegger's attention only after the statement of his initial position, as it evolved after the publication of *Being and Time.* The transition to Heidegger's later thought occurs through the so-called turning (*Kehre*), in which he maintained his concern with the problem of the meaning of Being, the so-called *Seinfrage,* even as he modified the approach to it. Almost everything about the turning, including its exact nature, location in Heidegger's writings, as well as its philosophic significance, is controversial.[9] Now it seems normal to expect that Heidegger modified his position in order to improve it. But since he also maintains that a true philosophy does not progress, the idea of the *Kehre* is problematic.[10] Either there is no change in his thought, in which case there is also no improvement, and the turning is only apparent; or there is a turning which improves the position, and Heidegger is inconsistent.

Third, there is a tension, perhaps even a basic contradiction, with respect to Heidegger's own notion of ontology. His question of the meaning of being, which he labels as ontology, is in fact an epistemological query, since his understanding of ontology is not the same as, say, the kinds of concerns which Aristotle raised in the *Meta-*

physics. Suffice it to say that Heidegger presents a modern view of metaphysics similar to that of Descartes, Kant, and Hegel, for whom metaphysics has become epistemology as opposed to ontology. But epistemology requires a concept of subjectivity, in Heidegger's terminology, *Dasein.* Hence, there is a tension in Heidegger's reliance on *Dasein* to pursue the study of ontology as he comprehends it and his desire to transcend a concept of the subject in order to let Being unveil itself directly without the mediation of *Dasein.*

Fourth, the discussion on technology provides a kind of litmus test for Heidegger's understanding of his relation to nazism. He turned to technology after a period as Rector of the University of Freiburg. Heidegger's explanation of his relation to national socialism after he resigned his post as rector is at least controversial. His conception of technology arises in the wake of his famous profession of faith, not in nazism as it occurred, but in the idea of nazism, what, in its revised form, he called "the inner truth and greatness of the movement."[11]

This statement provides an important indication of the presence, even in Heidegger's thought after the *Kehre*, of a continuing concern with the distinction between authenticity and inauthenticity, here applied to nazism. In order to reject other interpretations of national socialism as failing to grasp its essence, Heidegger commits himself to the capacity to differentiate the non-essential from the essential, in short to provide an authentic interpretation of national socialism. This statement further explains the later failure, even in the famous *Spiegel* interview, to distance himself from the holocaust. For to mention the holocaust, to acknowledge that national socialism was not incidentally but fundamentally linked to one of the most enormous crimes in the history of the world, would have required Heidegger to acknowledge that the possibility he saw in national socialism, namely as concerns the defense of the German university and the coming to fruition of the German people, was an enormous and deep error. Whether Heidegger was aware of this fact and desired to distance himself from national socialism as it occurred is an important question. Some have suggested that the discussion of technology belongs to a phase when Heidegger was trying to distance himself from nazism.[12] But if this was Heidegger's intention, he certainly did not succeed, since the discussion of technology is one of the less pellucid texts by this difficult thinker.[13] In fact, nothing in this text suggests that in writing this essay Heidegger was trying to take stock in any straightforward way of what had occurred and what his own role had been.

For historical reasons, Heidegger's view of technology is forever linked with his turn toward national socialism. The famous *Spiegel-Gespräch,* intentionally published only after Heidegger's death, and which was perhaps intended to provide Heidegger's own interpretation of his legacy, to function as a kind of testament, begins with a series of questions about Heidegger's relation to nazism. The interview then turns to a question about Heidegger's statement, made in 1935, about "the inner truth and greatness" of national socialism. In 1953 the lectures in which he made this notorious statement were published as the *Introduction to Metaphysics*. In this acclamation for the National Socialist movement, the words "namely with the meeting of planetarily limited technology and modern man" are placed in parentheses. In the interview, Heidegger defended himself against the accusation that he had only later added the words in parentheses, which, he claimed, were present in his original manuscript.

The interview then turns to Heidegger's view of technology. Several of his statements are worth noting here. He admits that he is unclear which political system fits modern technology and he states that he is not convinced that it is democracy. He states that human being is not in control of technology. And he asserts that at present philosophy cannot bring about a change and only a God can save us. He goes on to say that traditional thought, which he distinguishes from his own view under the heading of the other thought, is unable to comprehend the technological age. For Heidegger, the task of this other form of thought is to aid mankind in bringing about a satisfactory relationship to the essence of technology.

These comments on technology and national socialism bring together two elements in Heidegger's thought that are not entirely clear, and whose relation is, to say the least, troubling. In the *Spiegel* interview Heidegger suggests that his statement about inner greatness and the movement has been misunderstood and that it is actually linked to his own desire to do battle against modern technology. He argues that we cannot even think technology with ordinary philosophical thought, although such thinking is possible from the perspective of his so-called other thought which lies beyond metaphysical thinking. And he expresses doubts that democracy is relevant at this stage in the development of mankind.

For my purposes here, I will confine my comments to an initial discussion of Heidegger's view of the nature, or to use his term, the essence of technology. As an aid in this task, it will be useful to recall the development of the theme of technology in Heidegger's position. Technology is not an important theme in *Being and Time*. So far as I know, the only explicit mention of technology is in a single sentence in paragraph 69, which is a lengthy discussion of "The Temporality of Being-in-the-World and the Problem of the Transcendence of the World." In part b of this paragraph, called "The Temporal Meaning of the Way in Which Circumspective Concern Becomes Modified into the Theoretical Discovery of the Present-at-Hand Within-the-World," Heidegger writes: "Reading off the measurements which result from an experiment often requires a complicated 'technical' set-up for the experimental design."[14] But the basic conceptual framework within which Heidegger's analysis of technology will shortly emerge is already in place as early as this work in the famous discussion of equipment, where from a clearly pragmatic angle of vision he draws the well-known distinction between readiness-to-hand [*Zuhandenheit*] and presence-to-hand [*Vorhandenheit*]. He then says, in a reference to what nature is in itself:

Here, however, "Nature" is not to be understood as that which is just present-at-hand, nor as the *power of Nature*. The wood is a forest of timber, the mountain a quarry of rock, the river is water-power, the wind is wind "in the sails." As the "environment" is discovered, the "Nature" thus discovered is encountered too. If its kind of Being as ready-to-hand is disregarded, this "Nature" itself can be discovered and defined simply in its pure presence-to-hand. But when this happens, the Nature which "stirs and strives," which assails us and enthralls us as landscape, remains hidden. The botanists' plants are not the flowers of the hedgerow, the "source" which the geographer establishes for a river is not the "springhead in the dale."[15]

This passage already exhibits a claim characteristic of Heidegger's later writing, including his discussion of technology. In his account, there is an essence that surpasses, or lies beyond, his familiar distinction between readiness-to-hand and presence-to-hand. One instance of this essence is that of Being in general. Another instance is the essence of technology. Heidegger develops this idea further in his *Introduction to Metaphysics* (1935), where he equates Russia and America as two instances of the same domination of man by technology.[16] The important point is that the essence of technology, a still unknown essence, influences us greatly. This view is quickly expanded further in a series of later publications. For instance, in his two-volume study of Nietzsche, Heidegger discusses in passing the technological style of modern science and emphasizes the calculative reason of technology, a point which clearly goes back to Max Weber. And in the important essay entitled "The Age of the World Picture" (1938), he suggests that mechanization is the most visible extension of the essence of modern technology, which he describes as identical to modern metaphysics.[17]

The important point here is the link Heidegger tries to establish between technology and metaphysics. We know that his whole position is dedicated to the renewal of the problem of ontology, and we further know that for Heidegger this requires the overcoming of metaphysics. Since he here indicates, in reflecting on the nature of modernity, that machine technology is in some undefined sense identical to modern metaphysics, we can infer that for Heidegger the problem of technology is not merely an eccentric preoccupation but is instead central to his thought.

Heidegger develops his view of technology further in a short passage in the "Letter on Humanism" (1947). He comments on Marx's theory of alienation, which he describes as superior to other views because it attains an essential dimension of history. He then writes:

For such dialogue it is certainly also necessary to free oneself from naive notions about materialism, as well as from the cheap refutations that are supposed to counter it. The essence of materialism does not consist in the assertion that everything is simply matter but rather in a metaphysical determination according to which every being appears as the material of labor. The modern metaphysical essence of labor is anticipated in Hegel's *Phenomenol-*

ogy of Spirit as the self-establishing process of unconditioned production, which is the objectification of the actual through man experienced as subjectivity. The essence of materialism is concealed in the essence of technology, about which much has been written but little has been thought. Technology is in its essence a destiny within the history of Being and of the truth of Being, a truth that lies in oblivion. For technology does not go back to the techne of the Greeks in name only but derives historically and essentially from *techne* as a mode of *aletheuein*, a mode, that is, of rendering beings manifest. As a form of truth technology is grounded in the history of metaphysics, which is itself a distinctive and up to now the only perceptible phase of the history of Being.[18]

I have quoted this passage at length because it suggests important ideas which will reappear in his later discussions of technology. It also gives clear indications of the limits of his overall view. Heidegger presents his concept of technology in the context of remarks on the absence of a grasp of history in such other forms of phenomenology as those proposed by Husserl and Sartre. He prefers what he calls Marx's metaphysical view, one derived from Hegel. The superiority of Marx's position, Heidegger claims, lies in its superior understanding of history. This suggests that Heidegger accepts an adequate comprehension of history as a criterion for the evaluation of his own theory. But he rejects the efforts of Marx and Hegel to grasp history because they rely on the category of labor, which he describes as a metaphysical principle. Since he wants to surpass metaphysics, the theory of history he accepts must be non-metaphysical.

Heidegger follows a widespread presupposition in the secondary literature, strongly stressed by representatives of Marxism, of a difference in kind between idealism and materialism. Rejecting what he believes to be a metaphysical approach because it is metaphysical, Heidegger advances what he believes to be a third way that will surpass the split between Hegel and Marx, between idealism and materialism.

Marx regarded modern technology as something that arises within capitalism as a function of the process of capital accumulation. For Heidegger, technology is a metaphysical occurrence, whose essence has often been discussed but not often thought. In order to grasp the essence of metaphysics, he believes we need to connect technology to truth as found in the ancient Greek view of *techne*. He maintains that there is not merely an etymological link between *techne* and technology, but also an essential link; both are elements in the history of metaphysics, which itself belongs to the history of Being. Thus Heidegger argues that technology is an occurrence in the history of metaphysics and of Being which we need to comprehend from a perspective which attends to Being as the central theme, but from an angle of vision that surpasses metaphysics.

Now we must comprehend his understanding of what is at stake. He explains his view in a remark about why communism and Americanism are respectively more than a *Weltanschauung* or a lifestyle. According to Heidegger there is a relation between

metaphysics and the destiny of Europe; for Europe is falling behind in what he ob-
scurely refers to as the dawning of world destiny. Metaphysical thought is incapable of
getting a hold on this destiny. His claim can be summarized as follows: there is such
a thing as the destiny of the world, which is somehow related to Being; Europe has
failed to understand the destiny of the world because it approaches the problem
through metaphysical thinking. Heidegger offers an allegedly post-metaphysical thinking, said to be useful in enabling Europe to "catch up" by grasping an object of knowledge which cannot be known through metaphysical thought.

Clearly, Heidegger is making a radical claim for the relevance of his own form of
reason. Elsewhere he suggests that in the true sense thought exceeds both theory and
practice through its inconsequential nature.[19] This humble declaration only partly
masks the ambitious nature of his claim for thinking. Since the phenomenon in question is the essence of technology, we can infer that he holds that the fate of Europe, in
fact even world destiny, depends on the turn to a new, nonmetaphysical form of
thought alone capable of comprehending technology. For Heidegger, then, the idea of
technology is not a mere object of bewilderment, a philosophical conundrum, of
interest to the masters of abstract thought. Rather, philosophers, whether idealist, materialist, or Christian, are simply incapable of understanding technology. The understanding of technology is clearly fraught with a fateful consequence, since, in Heidegger's view, what is at stake is nothing less than the capacity of Europe to keep pace
within the fate of the world.

The main source of Heidegger's mature view of technology appears in a single
essay entitled "The Question Concerning Technology" (1954). Confining the discussion in this way is limiting, I am aware; a full discussion of all the material on Heidegger's view of technology would need to consider a variety of other writings. My present
aim is to come to grips with Heidegger's view of technology, in effect to think with him
on his own level in order to determine the outlines of his view of technology and to
evaluate it. In contrast to Heidegger, who did not believe in criticizing another thinker,
I see no reason to shrink from an evaluation of his (or any other) theory, no reason to
eschew an evaluation of his view of technology.

To a degree unusual even in Heidegger's writing, the discussion in "The Question
Concerning Technology" is murky. If, as Pierre Aubenque has claimed, Heidegger actually wanted to criticize national socialism in this essay, it seems reasonable that he
would have made a greater effort to be understood. The discussion is repetitive and
obscure, producing just the kind of hermeneutical issues which occupy scholars concerned (so to speak) more with the trees than with the forest, who in their attention to
such scholarly minutiae become unlikely to raise larger questions about the meaning
of his position.

Heidegger is constantly attentive to language, seeking to unveil what he regards
as original meanings that have been covered up over time. From that viewpoint it is

interesting that Heidegger entitles his discussion "Die Frage nach der Technik" and does not utilize the German term which literally means "technology." The word *Technologie*, which Heidegger does not employ here, derives from *Technik* and means roughly the knowledge of technique applied to a mode of production. The English term "technology" in the title of the essay translates *Technik*, but its closest equivalent in English is "technique," probably not "technology."

There is a difference in meaning here which is obscured by the translation. The English word means something like "the application of scientific knowledge to practical purposes in a particular field," or "a technical method of achieving a practical purpose." The German word means something like "the art of reaching a determined goal with the most purposeful and economical means," for instance the totality of means by which nature is rendered useful to man through knowledge and application of its laws. It includes the totality of tricks, rules, and procedures in a given domain, as well as forms of production, capacities, and aptitudes. Just as in German the term *Wissenschaft* is wider than the English "science," so there is an important difference between the domain covered by *Technik* and that implied by "technology." Whereas in English the field in view is something like applied science, in German it includes a wide range of fields such as construction work and even the theater. In short, "technology" is not an exact translation of *Technik*. This helps to account for a certain unexpected quality in Heidegger's account of technology which tries to elucidate a familiar phenomenon, namely technology, under the unusual heading of the term "technique."

Heidegger's discussion proceeds in his familiar apodictic style, mainly the absence of any recognizable form of argument. Since I think there is a considerable likelihood of going astray if we remain too close to the text, I shall try to restate what I regard as the central part of his view of technology in terminology more neutral than Heidegger's own language. In *Being and Time*, Heidegger posed what he called the question of the meaning of Being. In much the same way, the essay on technology raises the question of the meaning of technology, or more precisely the question of the essence of technology, where "essence" is understood in what he claims to be a nonmetaphysical sense. What he calls "essence" is in effect the essence of the conception of essence; it is something which endures, holds sway, administers itself, develops, and decays. In this connection, he calls attention to the etymological relation of the term essence [*Wesen*] to the verb *währen*.

But there is no easy way to justify or even to comprehend one strange statement: "Only what is granted endures. That which endures primally out of the earliest beginning is what grants."[20] This statement is strange since it seems to acknowledge that the essence is in some unacknowledged sense dependent on something beyond it. So in many religions, it would make sense to say that grass is the way it is because God made it, or that human being is made in the image of God. Heidegger seems to be suggesting that as part of his non-metaphysical understanding of the concept of essence, the essence is not to be understood in its own terms. To understand an essence we need to appeal to something beyond the essence, so that we can differentiate what is granted and, to employ consistent terminology, a grantor. The argument seems to rest on the relation of *Wesen* with *währen*, and of the latter verb with such further forms as *gewähren*, which, among other things, means "to grant." Further allusions to Goethe's usage of *fortgewähren* for *fortwähren*, although perhaps an interesting linguistic observation, cannot yield a conclusion in relation to the nature of essence. It is unclear why an essence needs or even could be granted, as if there were an initial agent which brought into being or maintained essences, such as Plato's demiurge. Heidegger seems to have something like this in mind in his concept of essence, but he neither clarifies nor supports this claim.

Just about the only unambiguous statement in the essay is Heidegger's passing comment that the essence of technology is ambiguous. The flow of his discussion indicates his concern to reveal the essence of technology and to differentiate technology from its essence. He notes that current concepts of technology are instrumental and anthropological, noting that technology is a means to an end but no mere means. Comments on the relation of instrumentality to causality lead to a further comment on revealing, which Heidegger relates to the concept of truth. He then asserts that technology is a way of revealing. Man does not control the unconcealment since modern technology is no human doing. A meditation on enframing—the horizon in which modern technology comes about—as the essence of the matter leads to further comments on destiny, danger, and the saving power.[21]

In this way, Heidegger insists on a profound difference between the utilitarian and essential aspects of technology. His basic claim is that technology is a way of revealing. This suggests that there is something beyond technology which is revealed by it. He also claims that the unconcealment is not under human control, that it is somehow associated with destiny, and that it is further linked to both danger and salvation.

The distinction between the utilitarian and essential aspects of technology at first glance appears relatively straightforward, even if Heidegger's statement of it is not pellucid. In more ordinary terminology, he appears to be pointing to the difference between utility and truth, which some pragmatists, notably James, are accused of conflating, and which Heidegger tried to keep apart in his differentiating of readiness-to-hand from presence-to-hand in *Being and Time*.

His further remarks about the essence of technology are less obvious, in fact often extremely difficult to construe. They can be reconstructed as follows: "Technology" derives from *techne*, which means "bringing forth," or *poiesis*, in art, or the fine arts, both in ancient Greece and, apparently by analogy, in technology. Now revealing entails that there is something concealed, as well as a means of unconcealment, in this case in technology. In this respect, the significant claim is that human being does not control the unconcealment, by which Heidegger understands the process in which the real shows itself in technology. For Heidegger, human being is confronted with a double incapacity: the presence of a realm of reality, perhaps Being as such, which is independent of us; and a process in which that which is independent can show itself at any time. Heidegger argues for this position by pointing to the fact that a forester is variously controlled by profitmaking in the lumber industry, the possibilities of cellulose, public opinion, etc. His point seems to be that the forester is passive with respect to the technology he employs. But this argument seems insufficient to justify an assertion that there is something like a realm of reality that is in fact revealed in technology; nor does it justify the inference that the revealing of technology itself is beyond human control. At best it shows only that it might be beyond the control of a given individual, such as the forester in the example.

It is worth noting that Heidegger locates various aspects of the technological process within social life. But at this crucial point in his discussion he fails to offer a theory of the social world. He needs to do this, however, if he hopes to understand the place of technology within human affairs. This step is also needed if he is to make good his claims that there is a further reality and that technology is beyond human control.

This lack is not a mere lacuna, an oversight; rather, it is a fundamental flaw. In defense of Heidegger, one might point out that he wants to go beyond the usual anthropological perspective in order to comprehend something deeper than or beyond the human dimension. But whether or not Heidegger can establish a transcendent dimension of technology, he still needs to locate technology in the social arena in which it arises and in which it functions. His manifest failure to do so is tantamount to a failure to analyze that aspect of the technological phenomenon which is closest to us and most familiar.

The central claim in Heidegger's view of technology is that its non-metaphysical essence lies in enframing [*Gestell*]. I will not reproduce the tortuous etymologies through which Heidegger suggests the meaning of this term. In any case, an etymological approach is insufficient to do the work Heidegger intends for it. There is no reason to believe that there are correct meanings which have been covered up in the course of time and which can later be uncovered in order to reveal the way things are in some deeper sense. Language is a matter of convention. What the Greeks meant by the use of a given term, which survives in later language, is only what they meant; it is not for that reason the true meaning of the later word. Earlier meanings are not true since language itself is not true; it is only an instrument with which to point, to refer, and to communicate.

Heidegger's concept of enframing is linked to his idea of standing reserve [*Bestand*, from *besthen*]. If we again recall his distinction in *Being and Time* between readiness-to-hand and presence-to-hand, the term *standing reserve* seems to mean something like the way things are disposed or ordered so that they can be utilized for various purposes. Heidegger suggests as much in his example of the plane on the runway ready for takeoff. But consistent with the concern in his later thought to surpass the anthropological approach, he does not want to identify purpose with human purpose. He does not argue for this view, but merely states it. One example is his remark that as a revealing, modern technology is no merely human doing. Another illustration of this quality in his writing is his sibylline description of enframing as "that challenging claim which gathers man thither to order the self-revealing as standing-reserve."[22]

The latter claim is especially significant, because Heidegger now supplements his earlier insistence on a trans-human reality and on technology as a mode of revealing with a new stress on the active role of that reality with respect to human being. An analogy would be Hegel's view of spirit which is said to realize itself in and through human actions. For Hegel, the subject of history is finally more than human. For Heidegger, similarly, what he refers to variously as reality or Being shows itself altogether independent of human being. But since Heidegger does not offer reasons for accepting a trans-human reality, it is rather difficult to accept that it is active with respect to human being.

The inability to sustain the claim that there is a trans-human reality which allegedly act through technology gravely compromises Heidegger's assertions about such aspects of enframing as destiny, danger, and saving power. His notion of destiny [*Geschick*] is not well articulated in his essay "The Question Concerning Technology." The concept reaches back in his thought at least as far as *Being and Time*. In that work he takes up the discussion of the "Basic Constitution of Historicality," and differentiates fate [*Schicksal*] from destiny [*Geschick*]. Heidegger describes destiny from the perspective of the resolute future-directedness he stresses there as "the historicizing in Being-with-others."[23] But it is notable that in *Being and Time* there is neither a substantial discussion of such phenomena as danger and saving power nor an association of either concept with his concept of destiny.

In the essay on technology, danger and saving power are associated with destiny, but neither the ideas themselves nor the purported relation is clearly elucidated. According to Heidegger, the destiny of revealing is not a particular danger, but danger as such.[24] But the nature of this general danger is at least unclear. Perhaps Heidegger has in mind death, which in *Being and Time* was described as the ultimate limit of life. If this is the case, it needs to be stated. And it further needs to be shown why revealing, or the destiny of revealing, is in general a danger, or dangerous. Conversely, Heidegger provides no ground for the association of danger with saving power other than a quotation from Hölderlin. But this is scarcely a sufficient reason.

In the present context, danger and saving power are secondary aspects; that is, they seem to depend upon destiny, more precisely upon the destiny of revealing. This leaves the question of the role of destiny in the phenomenon of technology unresolved. Looking back to *Being and Time* is helpful since the earlier discussion of historicality [*Geschehen*] led to the idea of a dimension of temporal happening which stands outside of resoluteness, even resolute future-directedness. Perhaps by analogy with Greek thought, Heidegger means to associate this with what the ancients called *moira*. But this is at most a piece in the puzzle. Once again Heidegger seems to be relying on associations between terms. So in German but not in English the terms *destiny* [*Geshick*] and *fate* [*Schicksal*] have an etymological relation to the verb "to send" [*schicken*]. From this association, we can jump to the conclusion that that which is revealed through destiny is sent and history is what is sent, or rather the process of sending. As he says: "We shall call that sending-that-gathers which first starts man upon a way of revealing, *destining*. It is from out of this destining that the essence of all history is determined."[25] If this reconstruction of his chain of reasoning is correct, what we have here is a change in his position. In *Being and Time*, Heidegger grounds historicality in time. Here he seems to be suggesting that the essence of all history (*Geschichte*) is based in a kind of sending, which leads to revealing.

This proposed reconstruction can be read in two very different ways. Heidegger seems to have in mind something like the Greek concept of actions beyond man's control, of that which just happens. This kind of claim is certainly admissible. Few writers care to argue that everything without limit is or could even conceivably be under the control of human being. For instance, in a famous passage Marx writes that men make their own history, but they do not do it as they please or under conditions they have chosen.[26] But if something is sent, then there must be a sender, which leads back by this complicated route to the transpersonal agent which Heidegger regards as operative in the historical sequence. This is what manifests itself through technology. This is a recurring claim, evident in his assertion that enframing acts upon man[27] and in similar assertions in his later writings.[28] But as far as I can determine, this claim is never argued but always merely asserted.

One of the tests of any philosophy which aspires to more than specialized competence in a restricted domain is its ability to comprehend the world in which we live. One of the important aspects of the modern world is the rise of modern technology. During the last hundred years or so the elaboration of applied science has enabled the application of new techniques to attain various goals both desirable and undesirable, but practically feasible within the existing social and political framework. Technology, then, is a means to an end which human beings determine and toward which, through this means, they strive.

If this understanding of the nature of technology is true, then Heidegger's view of the matter falls short of the mark in at least three ways. First, in his return to the Greek concept of *techne*, or art, Heidegger blurs beyond recognition the essential distinction between the concepts of art, technology, and technique. In his essay, he clearly has in mind technology in the English sense of the term, as his references to hydroelectric plants on the Rhine and aircraft clearly show. It is true that there is a poetic moment in technology, and that technology often requires using one or more techniques. To take a humble example, the technique of welding is applied in spotwelding, which is a form of technology utilized in various fields, such as plumbing. But despite the etymological associations upon which Heidegger plays, there is no reason to conflate art, technology, and technique.

It is helpful, indeed vital, to maintain the separation between a technique and a technology. Technology, like art, makes use of techniques, which are neutral with respect to both art and technology. If by "technology" we choose to mean something as general as the "practical implementation of intelligence," which in turn means that "the implementations are not wholly ends in themselves" but are "embodied in implements, artifacts, or sometimes in social organization,"[29] then we must keep apart the technology in general and its technical building blocks, or techniques.

If we take as our thesis the identification of art and technology, then it would be justified to argue that technology is no mere human invention but refers beyond itself. But since there is nothing other than a slim etymological link to point to a realm which technology reveals, we have only hortatory grounds to accept this conclusion. It may be appropriate to argue that genuine art is characterized by reference to a transcendent level of reality. But it is only on the basis of a deep conflation of art and technology that the latter can be taken for the former. This error follows directly from Heidegger's own terminology. For it is only by discussing technology under the heading of technique that he can discover a common term which links art and technology. To do so, however, is to mistake the nature of technology; for technology is not art, although some technology may have artistic results. Art is not technology, even if many forms of art routinely employ forms of technology.

In his attention to technology as the prolongation of metaphysics by other means, Heidegger fails to understand the way in which technology is embedded in modern social and historical contexts. On the one hand, his approach to technology through the problem of metaphysics sees technology not as an end in itself but as a means to a further end, in this case as part of his effort to destroy the history of metaphysics. But the terms *metaphysics* and *ontology* are not univocal. They have been understood in different ways in the history of the philosophical tradition. Aristotle, who is usually regarded as the father of metaphysics, did not use that term, which was only later applied to his writing. But he did speak of the science of being as being. Now it is clear why Heidegger draws a connection between metaphysics and technology. For metaphysics is connected to representational thinking, to what he elsewhere calls a world picture, in short to a two-worlds ontology which begins in Plato. Technology, as he understands it, presupposes dualism of this kind since its essence lies in enframing.

This explanation of how Heidegger arrives at his view of technology as an offshoot of metaphysics ought not to conceal the basically abstract, and therefore fundamentally incomplete, nature of his view. Whatever else it is, technology is not called into being through the problem of metaphysics. Heidegger's occasional suggestion that it is Being which lurks behind and affects technology is evidently mythological. For it is clear that technology, including modern technology, was called into being by human beings confronted with specific tasks arising within a specific social and historical milieu. There is no indication in "The Question Concerning Technology" or in Heidegger's other writings that he has a real comprehension of the nature of society or of its historical evolution. In *Being and Time*, where he was clearly under the influence of Kierkegaard, his understanding of the social as being-with was mainly negative, depicted as an inauthentic mode of life which one must leave behind. In later writings, there is no evidence that he has made progress in comprehending the social context. In this respect, he is clearly surpassed by Sartre, his most important French student. In his later writings, Sartre went beyond the abstract analysis he presented in *Being and Nothingness* to offer a concrete analysis of society in terms of such concepts as practice and lack.

Heidegger did not have a theory of society and, hence, cannot integrate his view of technology into a wider understanding of the social context. As a result, his theory of technology fails on at least two counts. First, he simply is unable to grasp the origins of technology from a historical perspective, which in turn leads him to attribute these origins to an extension of metaphysics. Second, he is unable to comprehend the social role of technology, which he also attributes to Being. It is clear that technology arose to satisfy human needs and desires. But Heidegger, who scrutinizes technology only within the context of his deeper concern with metaphysics, seems incapable of grasping the relation of technology to society and human being.

Because he cannot grasp the relationship of modern technology to the modern social world, Heidegger fails to comprehend the manner in which technology is not beyond human control but at least potentially subject to the wishes of human being. As soon as we admit that our actions often have consequences beyond our intentions, we can comprehend that there is something about history which escapes human control. But even if we acknowledge that in principle there will always be something about the historical process which will not merely coincide with the men and women whose actions constitute it, it does not follow that technology as such is beyond human control. That is a leap which no rational argument seems to require, but which Heidegger is apparently willing to make, because he has presupposed the action of a transpersonal reality like Being which, he believes, communicates with us through technology.

The deeper question is the extent to which technology is or can come under human control. I have already indicated why I hold that we must reject the claim that the technology we consciously employ for our own ends is something that lies under the control of a transpersonal reality. It is difficult to deny that technology sometimes seems to serve human beings. Obviously, there is an important distinction between one or another specific form of technology and technology in a general sense. Perhaps the best way to understand the idea of technology in general is as one aspect of modern social life. The history of mankind records the continued striving of people and groups to free themselves from natural and self-created constraints in order fully to develop their individual capacities. Whether human beings will finally be successful in this effort, whether we will ever free ourselves from even self-imposed constraints, is part of the larger experiment of human history whose outcome no one can know in advance. We cannot rationally assert, Heidegger notwithstanding, that man will never be able to bring technology under control. We may at least hope that such control will occur.

My conclusion is that Heidegger's reading of technology is deeply and irremediably flawed since he fails to identify basic elements of technology and conflates the phenomenon in general with technique and art. It is no accident that his reading of technology is flawed. The flaw follows directly from his profoundly antimodernist perspective. For Heidegger, the way to understand the world in which we live is through the elaboration of an alternative story of the meaning of Being. Such a story rejects the story mainly embraced since the pre-Socratics as one that arises from a false turning early in the path, a divergence from another, truer turning which would continue in the correct way. As in his analysis of modern thought, so in his analysis of modernity Heidegger proceeds from the assumption that what we think and what we are can be redeemed only by returning to an earlier and truer view of things. Hence, it is hardly surprising that he cannot grasp the way in which technology arose and functions within modernity since modernity itself, or at least what we mean by this term, functions for him as a fall away from an earlier and truer way of being, and perhaps even as a fall away from Being. I submit that neither Heidegger nor anyone else can comprehend modernity or its elements—including technology—unless these elements are approached in their own right without preconditions and allowed to tell their own story. Certainly, we are entitled to expect nothing less from any form of philosophy, including phenomenology.

The anti-democratic nature of Heidegger's thought is visible in his anti-anthropological conception of technology. As a political theory, democracy in all its forms necessarily presupposes a general conception of human being. In the course of the turning in his thought, or *Kehre* as he terms it, Heidegger turned away from the anthropological perspective which underlay his earlier approach to Being. His writings after the turning presuppose an anti-humanist perspective clearly described in his "Letter on Humanism." But a theory which goes beyond an anthropological concept is incompatible with any meaningful view of democracy which surely must depend on such a concept.[30] His anti-metaphysical theory of technology from the post-anthropological viewpoint of his new thinking is by definition non-democratic and even anti-democratic since it presupposes as its defining condition the rejection of the anthropological point of view which is the foundation of democracy. It is, then, no accident that Heidegger's fundamental ontology led him to embrace Nazi totalitarianism since his conception of Being is incompatible, and was understood by him to be incompatible, with any form of democratic theory.

Notes and References

1. See Victor Farias, *Heidegger et le nazisme* (Paris: Editions Verdier, 1987).

2. Lukács writes about Heidegger in different ways in a series of books. For the analysis of Heidegger's relation to Nazism, see especially *Existentialismus oder Marxismus?* (Berlin: Aufbau-Verlag, 1951), *passim* but particularly "Anhang: Heidegger Redivivus," pp. 161–83.

3. See, for instance, the recent, equivocal study by Jacques Derrida, *De l'Esprit* (Paris: Editions Galilée, 1987).

4. For this excessive judgment, see the discussion of Heidegger and Jaspers in *The Destruction of Reason*, trans. Peter Palmer (Atlantic Highlands, NJ: Humanities, 1981), chapter 4, part 6: "The Ash Wednesday of Parasitical Subjectivism," pp. 489–521.

5. See Martin Heidegger, *An Introduction to Metaphysics*, trans. Ralph Mannheim (New Haven: Yale University Press, 1977), p. 199. For the German-language original version of the statement, see Martin Heidegger, *Einführung in die Metaphysik*, in *Gesamtausgabe* (Frankfurt: Klostermann, 1983), volume 40, p. 208: "Was heute vollends als Philosophie des Nationalsozialismus herumgeboten wird, abermit der inneren Wahrheit und Grösse dieser Bewegung (nämlich mit der Begegnung de planetarisch bestimmten Technik und des neuzeitlichen Menschen) nicht das Geringste zu tun hat, das macht seine Fischzüge in diesen trüben Gewässern der 'Werte' und der 'Ganzheiten.' " For Heidegger's misleading effort to suggest that this statement had merely tactical value in order to defend freedom of speech, see p. 233, where in a letter to S. Zemach, in a comment on this passage, he writes: "Die verständigen Hörer dieser Vorlesung haben daher auch begriffen, wie der Satz zu verstehen sei. Nur die Spitzel der Partei, die—wie ich wusste—in meiner Vorlesung sassen, verstanden den Satz anders, sollten es auch. Man musste diesen Leuten hie und da einen Brocken zuwerfen, um sich die Freiheit der Lehre und Rede zu bewahren."

6. See *Existentialismus oder Marxismus?*, p. 19.

7. Ibid., p. 28.

8. See John D. Caputo, "Demythologizing Heidegger: *Aletheia* and the History of Being," *Review of Metaphysics* 41 (March 1988): 542.

9. The best treatment of this problem of which I am aware is Jean Grodin, *Le Tournant dans la pensée de Martin Heidegger* (Paris: Presses Universitaires de France, 1987).

10. "Letter on Humanism," in Martin Heidegger, *Basic Writings*, ed. David Farrell Krell (New York: Harper and Row, 1977), p. 215.

11. See Heidegger, *Introduction to Metaphysics*, p. 199.

12. See Pierre Aubenque, "Encore Heidegger et le nazisme," *Le Débat* 48 (January-February 1988): 120.

13. See "The Question Concerning Technology," in Heidegger, *The Question Concerning Technology and Other Essays*, trans. William Lovitt (New York: Harper and Row, 1977).

14. Martin Heidegger, *Being and Time*, trans. Macquarrie and Robinson (New York: Harper and Row, 1962), p. 409; *Sein und Zeit* (Tübingen: Niemeyer Verlag, 1963), p. 358.

15. Heidegger, *Being and Time*, p. 100; *Sein und Zeit*, p. 70.

16. Heidegger, *Introduction to Metaphysics*, p. 38.

17. Heidegger, *Question Concerning Technology*, p. 116.

18. Heidegger, *Basic Writings*, p. 220.

19. Heidegger, "Letter on Humanism," in *Basic Writings*, p. 239.

20. Heidegger, *Question*, p. 31.

21. The clearest statement of Heidegger's view of enframing of which I am aware is his characterization of *Machenschaft* as *Der Bezug der Unbezüglichkeit*. This expression can be rendered very approximately as the unframed frame—in short, as a notion of horizon. See Heidegger, *Beiträge zur Philosophie (das Ereignis)*, ed. F. von Hermann (Frankfurt: Klostermann, 1989).

22. Heidegger, *Question*, p. 19.

23. Heidegger, *Being and Time*, p. 438; *Sein und Zeit*, p. 386.

24. Heidegger, *Question*, p. 26.

25. Ibid., p. 24.

26. See Karl Marx, "The Eighteenth Brumaire of Louis Bonaparte," in Marx and Friedrich Engels, *Basic Writings on Politics and Philosophy*, ed. L. Feuer (Garden City, NY: Doubleday, 1959), p. 320.

27. Heidegger, *Question*, p. 24.

28. See the denial that man controls technology in Heidegger, *Identity and Difference*, trans. J. Stambaugh (New York: Harper and Row, 1969), p. 34.

29. See Frederick Ferré, *Philosophy of Technology* (Englewood Cliffs, N.J.: Prentice-Hall, 1988), p. 26.

30. Heidegger emphasizes the anti-democratic nature of his thought in the *Spiegel* interview in the famous claim that only a God can save us. See "Only a God Can Save Us: Der Spiegel's Interview with Martin Heidegger," *Philosophy Today* (Winter 1976). For an earlier version of this claim as the idea that only *das Deutsche* can save us, see Heidegger, *Heraklit*, ed. M. Frings (Frankfurt: Klostermann, 1987), pp. 107–108 and 123.

PART IV

Media Theories: The Politics of Seeing

THERE IS IN Heidegger a nostalgia for a more authentic and immediate experience than the technological appropriation of being. It is true that modern experience is increasingly mediated by technical systems and devices. This trend, which first makes its appearance in the sciences, takes on enormous importance in the democratic era. These mediations are not neutral; they have a logic of their own and shape the content of the experience they mediate. The two chapters in this section explore the role of new ways of seeing in modern times, and evoke the postmodern consciousness of total mediation, beyond any nostalgia for immediacy.

Don Ihde's "Image Technologies and Traditional Culture" argues that modern experience has been fundamentally altered by new technologies of representation. Just as the microscope and the telescope reshaped the scientific view of the world, so do everyday image technologies such as cinema and television reshape the social world. The erosion of traditional cultures' self-certainty is especially significant. Defensive movements such as Islamic fundamentalism attempt to hold back the invading images of modernity. But the future promises neither a return to traditional culture nor a continuation of modernity as we have known it up to now. Rather, we are headed toward a world "pluriculture," based on universal access to a *bricolage* of cultural fragments drawn from all over the globe.

Modern culture is influenced not just by media technologies but also by the epistemologically bestowed privilege of sight as the surest basis of objective knowledge. That epistemological bias assigns the new media their role in the public processes of modern societies.

This is the argument of Yaron Ezrahi's chapter, "Technology and the Civil Epistemology of Democracy." Like Rousseau's "civil religion," Ezrahi's "civil epistemology" refers to essential dogmas underlying modern democracy. No democratic order is possible without the belief that political truth can be made manifest for all to witness. The rise of that assumption is related to parallel changes in the sciences. Just as Boyle's air pump served to produce publicly assertainable evidence for his theories about the vacuum, so "democratic fictions of the real" are produced by the communications media to define public actions and assign responsibilities. However, the postmodern sensibility is sceptical of these productions and so undermines the civil epistemology of democracy.

9

Image Technologies and Traditional Culture

Don Ihde

A T THE FOREFRONT of much scientific but also of communication development
stands a wide variety of "image technologies." In the scientific domain many of
these technologies are designed to portray *interiors* without destructive or invasive
procedures, such as CAT scans, MRS imaging, and sonograms, or to investigate extreme macro- or micro-structures of natural entities, such as computer-enhanced photography or electron microscopy.

Image technologies within science *return observation to a perceptual dimension*
which remains one of the primary valued trajectories in a largely visualist emphasis
within science. Moreover, such perceptual visualism has all the advantages of immediacy, pattern recognition, and gestalt quality which characterizes the informed style
of vision appropriate to scientific observation. Such developments, although far more
complex and intricate, continue what Derek de Solla Price called the "artificial revelation" which Galileo's telescope began.[1]

Microscopes, telescopes, X-ray imaging, and now the above noted contemporary
examples belong to a long history of perception-enhancing technologies which *embody* scientific observation. They are an essential element in the *technological embodiment of science in its instrumentation.*[2]

Accompanying this technological trajectory, and continuing the Galilean sense of
discovery, there remains a fascination with what is shown, particularly when it is novel
or previously unknown. We retain a sense of awe with respect to the latest depictions
of forms of galaxies—spiral, crab, etc.—or with our newly attained ability to image
atoms.[3] And, as with Galileo, what is revealed within the aura of fascination tends to
override what is often less clear, the dimension of what is *concealed* or reduced by the
instrument-enhanced vision.

While modernity clearly sides with Galileo, no longer doubting the mountains of
the moon, the rings of Saturn, or the satellites of the planets, the Churchmen of his time
did have a point in that early telescopy also did have almost equally obvious limitations. Indeed, the much later parallel in which there was strong resistance to microscopy in biology was motivated not so much by religious reservations as by the awareness that "reading" the "images" contained far too much ambiguity and subjectivity

(until aniline dyes helped create stronger contrasts in the specimens—not unlike today's use of "false color").

One could almost say that with both the more primitive technologies of the telescope and microscope there was a possible gestalt evaluation related to the ratio of revealing/concealing within instrumentally embodied vision. Defenders could rightfully note that "something" was being revealed, and that the macro- or micro-reality so shown was never seen before and therefore the device was a new means of discovery, while detractors could emphasize the concealment factors which should have led to critical doubts about "what" such entities could be. Indeed, the scenes were often complicated by objects apparently "seen" which were only later discovered to be effects of the instrument itself—instrumental artifacts.[4] Yet, in the end, the defenders have prevailed and there are today means of critique which in the experimental contexts include elaborate means of differentiating instrumental artifacts from observed phenomena. In the context of large instrument experimentation, such processes may be exceedingly difficult to determine and arguments about where and what is seen may proceed for years.[5]

This same history and process, however, also belongs to a much more public and *cultural* domain. Image technologies belong as intricately to our contemporary ordinary lifeworld, and *embody it in ways not dissimilar to the technological embodiment of science with respect to cultural dimensions.* The obvious equivalents to science's instrumentarium of image technologies in cultural life are the cinema, television, the growing spectrum of perception-enhancing and reproducing devices such as camcorders, VCRs, cassette players, CDs, etc.

The implication, in large, is also clear: what are the lifeworld equivalents to revealment/concealment? to instrumental artifacts? to trajectories of development? It is at this point that the central theses of this chapter may be entered:

1. Modern science, since the Renaissance, became technologically embodied in its instrumentarium. Its perceptions are often instrumentally transformed. In parallel fashion, contemporary culture in its now global communications context is increasingly embodied through its instrumentarium, again image technologies.
2. One may argue that not only was the irreversibility of a technologically transformed vision in science a major factor in scientific "seeing," but that its effects were *acidic* to previous modes of unmediated perception. The whole realms of micro- and macro-entities revealed instrumentally overcame the narrower realities of unmediated perceptions, so that now mediated perception is fully sedimented within the scientific community. Here, within contemporary culture, one may argue that image or communications technologies are similarly acidic to prior modes of traditional cultures.

This investigation is an inquiry into the ways in which "image technologies" are non-neutrally acidic to all traditional cultures.

A Phenomenology of Technological Imagery

The first step in establishing the non-neutrality of image technologies with respect to traditional cultures will be a brief phenomenology of some of the structural features of technologically transformed perceptions. However, phenomenologically speaking, an initial qualification must be noted with respect to the prevailing terminology regarding "images" and their relation to "representability."

In a sense, "image" and "representation" belong to an implicitly Platonistic tradition within Western philosophy. Their use often implies that there is some "real" foundation, of which there are layers of secondary or tertiary "imitations" or "representations" and around which one may then generate any number of "realist/anti-realist" arguments. Phenomenologically the very term *image* becomes questionable. The so-called image itself is a *phenomenon*, that which can be seen. But what is seen must be analyzed in terms of its own context and field, within which it presents itself—it has its own unique mode of positive presentation. And, as will be shown, this very factor is one of the most subtle in the phenomenology of "imagery" itself.

I shall begin with an analysis of a few rather larger features of image transformations by contrasting mediated with unmediated perceptions of objects. Return to the transformations implicit in the Galilean use of a primitive telescope: telescopic vision is *framed vision*. Seeing through a tube changes the field within which the object is seen. The surrounding presence of the tube *changes the context* of the object seen.

Historically, although it was a complex instrument of compound lenses, even seeing *through* Galileo's primitive telescope was difficult. It has been likened to seeing a distant target through two small keyholes a yard apart! But, once targeted, the moon becomes framed within the limits of the telescope's tube. The "image"-moon is removed from its ordinary field within the sky and is framed within a now limited instrumental "field."

This transformation contains much, much more as well. The entirety of "apparent" space-time is also transformed. Relative motion, which is an indicator of space-time, becomes magnified as much as the moon as object. In a hand-held telescope it is very hard to "fix" one's visual object and bodily movement—jerks, inability to follow the now enhanced movement of the moon through the sky, coordination of eye-hand motions—all intrude into a previously already learned and "natural" seeing.

Spatially the moon is magnified as well. But it is magnified in the new context of difficult-to-control bodily motion. Both body and moon are thus magnified. And both factors are part of the now technologically transformed observational context. One may note, reflectively, that part of what occurs in the transformation is an implicit dialectic between the previously already sedimented form of unmediated seeing and the new, as yet to be fully appropriated form of mediated seeing.

Third, in the selective magnification/reduction of telescopic seeing, "apparent" distance is also transformed. The body-moon distance is "diminished"—relativistically it is equivalent to say that one's body is now "apparently" closer to the moon or that the moon is "apparently" closer to the viewer. But, more subtly, note that the very use of "apparent" implies at least a dialectic between technologically mediated vision and ordinary vision, or, at most, the implication that there is a primacy to the ordinary position.

Time dimensions are also implied in this transformation. The magnification of minute motion in the mediated situation also magnifies instantaneous aspects of time not likely to be noted in ordinary contexts. What was developed later in the history of telescopy—motorized mechanisms which allowed telescopes to "follow" the earth-celestial body motions and thus "fix" the object for continuous "fixed" viewing—was an attempt to reduce this time magnification.

Although I shall not here trace out all the implications to be noted in this preliminary phenomenology, it is interesting to note that much of the trajectory of development of the telescope revolved around increasing the *clarity* of the "image" (and thus the "transparency" of the instrumental capacity) as well as *producing* a fixed and repeatable object as the aim of the observation. In principle, such a fixing is the equivalent of creating a *photograph*.

Photography

The celestial object "fixed" in repeatable time and realistically "represented" is simultaneously an idealized photograph and implicitly reveals one goal of an experimental value. In its actual history, photography had a long preparation, but once invented it also led to a number of interesting cultural transformations.

As early as the Renaissance, the *camera obscura* had become an object of fascination. It embodied a kind of linear perspective in its inverted images, and according to Lee Bailey, later even became a major metaphor for the modern subject in Descartes.[6] Photography, later, allowed the *fixing* of the same kind of image found in the *camera obscura*. Photography itself was a nineteenth-century technology with the development of the daguerreotype in 1839. Its invention was an immediate success followed by runs on optical shops for lenses for cameras, by hysteria from painters, and by fascination with the photographic representations which were produced. It was an immediate popular cultural technology.[7]

Phenomenologically, photography does many things noted previously in the case of the Galilean telescope, but also with differences: (1) it presents its "image" framed, i.e., it abstracts out of the living contextual field a specially focused and selected and narrowed image field; (2) in this case, rather than magnifying the body-object viewed motion, it *freezes* or *fixes* apparent motion, making it "still"; and (3) its result is a striking *isomorphism* or apparent "realism" between its image and the thing "imaged."

I shall not return here to the already interesting studies which show that both (1) and (2) are learned modes of seeing which, while already prepared for and thus more invisible in Western Euro-American cultures, were more obviously in need of sedimentation for non-Western peoples. Instead, I shall take note of a few salient features in the mode of "representation" which followed the development of photography.

Using historical representations as a kind of phenomenological variation, recall the early pre-photographic representations of Amerindians by the first European artists in the Columbian era: Although the subjects are depicted as nude or with scant garments and feathers, the physiognomies as drawn do not differ much from standard Europeans. Conversely, when Perry arrived in Japan as late as the nineteenth century, local Japanese artists portrayed the Americans as very Oriental in appearance, again apart from dress which did more accurately represent the Western uniform. Both these cross-cultural examples show us something about a kind of naive viewing which is very unphotographic.

Similarly, while the history of—particularly European—painting did illustrate a trajectory toward "realism," viz. Dutch painting in the seventeenth century, it remained highly constructed and could not be compared to the post-photographic "photo-realist" style of the present. Interestingly, the "realistic" paintings of Amerindians in the nineteenth century by Catlin and Remington are also post-photographic. *The photograph, more than merely representing, "teaches" a way of seeing.*

Its "realism," however retains within itself a distinct difference between it and what it represents: the very fix which fixes space-time, often metaphorically called "dead," gives the permanence of the photograph its own kind of *irrealism*. Its immobility, however close to the Greek or even scientific ideal of a "fixed truth," remains *only* that which is "represented." Adding motion to the picture radically changes that variable—the "moving picture."

Cinema and Its Clones

Motion pictures followed still photography by fifty years (1889, with "Fred Ott's Sneeze"). And by adding motion to the previously still or space-time fixed representation, gave a new sense of life-like realism. The famous "L'Arrivée d'un train en gare" of 1895 showed from a low angle, close up, a steam train so realistically approaching the early cinema audience that they would cry out or move. Cinema—and now television and other audiovideo forms—has obviously gone even further than the "realism" of the early documentary. And one important effect is the lowering of the "realism/irrealism" distinction still more strongly apparent in the photograph.

I shall turn to two very recent events to heighten this effect: First, the Vietnam War of the sixties brought a kind of "television realism" into virtually every household of the West. Then under the freer ranging rules covering journalists (still derived from World War II), the battlefield and even civilian casualties were nightly brought before a populace at large. One effect was clearly repugnance at war, particularly that kind of war. By the 1990 Gulf War the "rules" for journalists had been radically changed.

The news was clearly much more pre-selected. "Smart bombs" were repeatedly shown flying accurately into ventilators or hitting only focused targets (implicitly indicating that indiscriminate bombing, civilian casualties, etc., were being minimized), nighttime successful Patriot defenses against Scuds, etc.

The "real footage" carries with it an implied realism, even though it was the *same* footage which was repeatedly shown day after day. It is the implied realism of the "documentary" which carries this ombra and is thus taken as real. It has taken until 1992 for the critical responses to begin to show what was *concealed* through the apparent realism of the newsbroadcast, i.e., actual numbers of military compared to civilian deaths, low percentages of successful Patriot intercepts, offensive events like the live burial of Iraqi soldiers, etc. But what was most fascinating was the simultaneous *immediate awareness by the viewers that the news was being "cooked" and yet its planned effect was accepted and even celebrated.* It was as if the "realism/irrealism" of television was itself the new "real thing."

The same realism/irrealism phenomenon occurred with respect to the 1992 U.S. presidential election campaign. Morality debates now occur over depictions which occur in soap operas as though they were real events (single mothers giving birth, teenagers being initiated into sexual activity), even though the viewers are fully aware of the presumed difference between the representation and the represented. The *medium* contains in its realism/irrealism presentation an effect in which "image technology"-mediated and unmediated ordinary experience again becomes dialectical. What could be missed here is the parallel to the previously noted history of science examples. Although in the early days both telescopes and microscopes fell under severe criticism as means of gathering knowledge, with time both were sedimented into ordinary scientific consciousness in such a way that they became taken for granted.

Phenomenology Redux

In the examples heretofore, I have taken early optical technologies in science and somewhat anachronistically compared them to more contemporary cinematographic (and its clones) technologies. But there are special, artifactual aspects to cinema-like image technologies which should now be noted as well:

1. Some effects which dialectically contrast between visualization in ordinary sight and image-mediated sight have long been known. Seeing through lenses (a) diminishes depth. The object field is "flattened." This is particularly dramatic with television in which, contrary to Cartesian or linear perspective space, the size of the figure in the background remains the "same" as the figure in the foreground (viz. the pitcher on the field compared to the batter in baseball). (b) Contrasts are often different, thus the need today to enhance contrast, or even to move to "false colors." (c) At the gestalt level, reproductions are never entirely isomorphic with the objects reproduced—even

today's digital enhancements which create "hyper" real images are noticeably different from the actual objects.

2. But in the case of cinematographic-like technologies, it is the space-time *constructive* aspects which often stand out. Time reversals, flashbacks, special effects, and, above all, discontinuities (very pronounced in MTV presentations) have been raised to a high technical art. Ordinary space-time is here technologically deconstructed and reconstructed in a *bricolage* of image space-time.

While it may be initially obvious that all these phenomena go into the transformation of contemporary cultural perception, it may not yet be so obvious how such phenomena non-neutrally impact upon traditional cultures.

Revivals of Traditional Culture

If contemporaneity contains a multiplicity of strands, including what is often called the "postmodern" and the conflict between modernity and the traditional, then the revival of strong forms of traditional culture at present may seem to be quite anomalous. Three such movements may be noted for examples: (1) The strongest such traditionalist movement is the revival of Islamic fundamentalism (although other fundamentalisms are also waxing); (2) if Eastern European communism was initially viewed as a "modernization" process, its collapse in the last decade produced in its wake ethnic and nationalist revivals and these currently characterize much of the strife in Eastern Europe; and (3) most prominently in the U.S.A. may be noted a kind of "neo-Puritanism" which is broader than long indigenous forms of religious fundamentalism or political rightism.

All of these movements are virulent, prominent, and apparently growing. It thus might seem that, with respect to traditional cultures, contemporary image technologies are producing results contrary to the thesis of non-neutral acidity.

Or, could it be that there is a deeper parallel than meets the eye in the very re-emergence of virulent traditionalism within the context of contemporary image technology? Were the thesis that parallels early modern science with its instrumentarium again followed, one could at least note that Galileo himself lived in the very context of the Inquisition which historically was a response to both the rise of Protestantism and the expansion of colonialism (and a contact with the new cultures) in the New World.

There are clear aspects of all three of the above noted revivals which do appear to be *reactions* to what is often conveyed through image technologies: (1) One often-noted factor in the collapse of Eastern European communism was the frustration over the lack of consumer goods in those countries. Related to this frustration was the portrayal of Western European and American societies which were rich in consumer goods—often the contrast was conveyed by image technologies, particularly movies and television. (2) Islamic societies often appear caught between conflicting responses to modernity. Does one modernize moderately by adapting high technology, but retain some core of Islamic culture? Does one lead a double life in which one is thoroughly Islamic at home, but thoroughly Western abroad? Or, does one react to modernization by reviving the religious pride implied in Islamic fundamentalism? In the cases of the more conservative Islamic movements what is of particular interest in this context is the contrast between the overt willingness to adopt totally contemporary weaponry and seek modern forms of productivity (oil production technologies, for example) while simultaneously censoring as fully as possible the image media from magazines to television. (3) In the case of American neo-Puritanism the role of image media is clearly prominent. Public radio is under attack for its presumed "liberal" tendency; the already noted soap-opera morality debates point to neo-Puritanical pressures upon television; and, as part of the same phenomenon of reaction, the attempt to accumulate more and more of the media channels by religious—primarily evangelical or fundamentalist—entrepreneurs, all belong to a response to the impact of image technologies.

Image Pluriculture

An indexical question to open the way to the display of image non-neutrality is one which asks what is "imaged," and how that which is imaged occurs, which stimulates the various revivals or reactions of traditional culture.

Clearly the contemporary context is one of *global communication networks*. The image technologies occur in different ways in almost all the cultural contexts of the world. In countries in which television is seen as a tool to "leapfrog" from tribal village life into the twenty-first century (India, for example), satellites beam programs into central village halls in which a single television set is watched by all. In more developed countries television plus many other communications technologies are decentralized into homes and offices.

What is shown? Often, an intellectual's temptation is to judge content in terms of a conflict between high culture and *popular* culture. It has been observed that "Dallas," an American soap opera, was perhaps the most universal of all programs in the last decade. In 1984 when I last lived in France, one topic of conversation among media personnel was how they could create a successful indigenous counterpart to retain their own French flavor! One can easily see that defensive cultures can immediately raise the question of program selection, control, or censorship. During the height of the Bhutto era in Pakistan the single most burning issue with respect to public television was how to keep Western portrayals of women off the television. But it is by taking a virtual "postmodern" example from highly developed countries that one can best see what is being delivered.

What is being delivered is *"pluriculture,"* a unique image-technological portrayal of multiple "otherness." Any international news program portrays a spectrum of countries and cultures in a single broadcast (Iran, Iraq, the Sudan, various European and American, Japan, India, South Africa, etc.). And, to take a most extreme counterexample from popular culture, MTV presents the most fragmented set of images from a multicultural mix of musics, fashions, ethnic traditions, and human races. Both the news and MTV are daily pluricultural *bricolages.*

The esoteric, the novel, the pluricultural—perhaps ironically—is not too different from Galileo's moon mountains, planetary rings, and unknown satellites. A tribal definition in which one's own group could be the only true human beings becomes increasingly difficult to maintain in a context in which the "others" are present in one's own living room.

But the "other" can also be a threat. Implicit in pluriculture is a kind of *bricolage* relativism. One may pick and choose culture fragments, multiply choices, and in the process reflectively find one's own standards often provincial or arbitrary—certainly no longer simply a priori obvious. The pluricultural presence, moreover, characterizes not only the news or MTV but many kinds of programming. In one sense, of course, it is an actual reflection of a global experience, albeit mediated through image technologies which are the instruments by which pluriculturality is initially and sometimes solely viewed.

The parallel with the Galilean telescope is close—no human so far has actually viewed Mars firsthand, any more than most of us have lived among the Berber, but both viewings are now part of our technologically mediated experience.

If the content is often pluricultural, the *form* should not be overlooked. The capacity of image technologies to fragment, deconstruct, only to reconstruct in a series of often juxtaposed, non-linear imagery, *re-enforces* the *bricolage* quality of what is shown, whether it be the short news item or the fleeting image of Madonna in a pose on MTV.

Although the above is obviously selective, it is also indicative. The portrayals on both the news and MTV stand in stark contrast to more traditional depictions of narratives, logics, progressions which are recognizable now as a more archaic form of broadcasting. The "talking head" of the academic television lecturer is simply received by most as boring. The animated talk show is a step up, but it is superseded by the forms which display the irreal realism of televisual "being there" in the on-scene camera or edited video. It is time to get "behind the scene."

Edited Reality

How is the evening news or the MTV video constituted? In the case of the newsroom, the editing is done from the spectrum of multi-screens which display, simultaneously, the muliplicity of events going on in the world (these already selected from a vast variety of sources). The technician-editor scans the compound screens and selects and moves from one to another, forming a discontinuous narrative of news fragments. The result is a "constructed" result, fragmented, but deliberately designed.

Similarly, the video is spliced and constructed from shots and takes, remnants from the cinema, united by the narrative of the single song being sung over the discontinuities of the images. The "reality" is a constructed, edited reality.

I wish to note two aspects in this construction by editing. (1) The content which is edited is already the diverse set of multiple phenomena. The multiple screens from which the news is constructed refer to an ever-increasing "tree" of possibilities which may be shown. But (2) the viewing which edits is itself a kind of compound vision. It scans and selects from among the fragments, itself a fast-moving process which must fit the style of the medium. (In fact, academicians should actually recognize a similar process in the now dominant contemporary mode of composition! Most of us use word processors which also mimic the same factors. Some experienced editors have noted that not only are manuscripts longer now—due to the ease with which footnotes, re-arrangements, and other composition may occur—but are often marked by stylistic features which indicate that "text bits" may have been moved around in the composition-editing process speeded by word processing.)

What has been described are factors related to image technologies in operation. Such technologies also display "instrumental inclinations": cinema-like image technologies could, of course, restrict themselves to reproducing simply speech, but the inclination is usually more actional. That which is dramatic, reducible to the shorter bit, novel, even esoteric, is more likely to be watched. And even when what is displayed is mundane—such as a Charles Kuralt travelogue through heartland America—it is made to be displayed as curious or novel. Image technologies "anthropologize" the natives.

In the inclination, a very interesting parallel again appears between contemporary communications phenomena and the introduction of instrumentation into early science. What instrumental mediations do is let one see what was previously unseeable, particularly the micro- or macro-phenomena which eventually became the daily fare of science frontiers. What is fascinating is not simply that which is seen in ordinary vision. A counterexample from the history of art is instructive: one could have gone around the environment and simply held up a frame filled with plate glass if the point were simply to mediate technologically what is seen—but it is the *transformation* which occurs through variational change, magnification for example, which makes the newly mediated vision interesting. In the cultural counterparts a similar "magnification" phenomenon occurs.

Intellectuals deplore, in the popular media, the extravagant displays of sex and violence which become the common fare. Interestingly, this selectivity is a side effect of an image trajectory. It magnifies what is initially titillating, escalates it into a kind of image-mediated frontier for entertainment observation. The kind of sex and violence portrayed does not characterize most mundane life or ordinary perception—but neither do the micro- or macro-entities of today's frontier science characterize most mundane life or ordinary perception. So, although in radically different contexts, the *magnification-reduction phenomenon* shows the same thing. And in both domains the transformation phenomenon is acidic to traditional forms of seeing. Admittedly, our recent fascination with "buckyballs" may seem very different from the extremities of *Terminator II*, but both are magnifications beyond the mundane made possible through image technologies.

Life Imitates Art

Galileo, in effect, created a new "art" through his telescope, a new way of seeing. In science such a way of seeing became so thoroughly constituted that today one may say that the carpentered products of science—perhaps more, or at least as much "technologically constituted" as "socially constructed"—take both their shape and measurement in terms of science's technologies.

In the ordinary lifeworld, image technologies produce, deconstruct, and reconstruct in a real/irreal artform ways of seeing on the cultural plane. The results, again not unlike the history of science, are mixed. What is unarguable, however, is that the image mediation has made present a kind of irreversible awareness of pluriculturality which challenges not simply the groups which today strongly react, but all our cultural forms. The beginnings of a questioning of Euro-centrism are first responses to what will become a much stronger transformation of awareness so long as we maintain the mode of image mediation which now obtains.

Notes and References

1. See Derek de Solla Price, "Notes Towards a Philosophy of Science/Technology Interaction," in *The Nature of Technological Knowledge* (Dordrecht: D. Reidel Publishing Co., 1984).

2. See my *Instrumental Realism* (Bloomington: Indiana University Press, 1991) for a more complete account of an array of philosophers of science on instrumentation.

3. In a typical science-industry occurrence, the first published pictures of atoms were arrangements of the logo "IBM."

4. Ian Hacking in his *Representing and Intervening* (Cambridge: Cambridge University Press, 1983) has an excellent discussion of the history of the microscope; see pp. 186–209.

5. A recent history of a large instrument experiment and a discussion of the problem of in-

strumental artifacts may be found in Peter Galison, *How Experiments End* (Chicago: University of Chicago Press, 1987).

6. Lee W. Bailey, "Skull's Darkroom: The *Camera Obscura* and Subjectivity," *Philosophy of Technology* (Dordrecht: Kluwer Academic Publishers, 1989).

7. H. Gernsheim, *The History of Photography* (New York: Oxford University Press, 1955).

10

Technology and the
Civil Epistemology of Democracy

Yaron Ezrahi

Introduction

DEMOCRATIZATION IN THE West, and particularly in its American variant, reflects a move to externalize political power and redefine politics as a human enterprise unfolding in the visual field of common perception.[1] In contrast to the evident theatricality of the monarchy and the splendor and pomp of the court as an ongoing political spectacle, the body politic of democracy was presented as honestly transparent to the people. "A Constitution," wrote Thomas Paine, "is not a thing in name only but in fact. It is not an idol, but real existence; and wherever it cannot be produced in visible form there is none."[2] Democracy, Paine insisted, is a kind of government which "presents itself in the open theater of the world in fair and manly manner. Whatever are its excellences or its defects, they are visible to all. It exists not by fraud or mystery."[3] Paine's statement aptly captures the Enlightenment trust in the powers of light and visibility to demystify political power, render the exposed political actors more honestly accountable, and enlist the sense of sight as a vehicle of democratic political participation. It is in this context that technology, including machines and machine metaphors, has come to play such an important role in the particular democratic genre of public action as a political spectacle, and more generally in the very construction of democratic politics.

Underlying this development is, of course, a deeper process through which the burden of creating and sustaining the sociopolitical order has shifted from divine agencies, where it was largely presumed to be located in the premodern state, to human agencies. While the works of divine agencies appeared manifest in nature and in the rise and fall of whole societies, politics and technology have become principal theaters of human historical enterprises.

This transformation has been associated with the decline of mythical-ritualistic conceptions of the exercise of human power and their replacement by more instrumental-causal notions of human intervention in constructing or mending the sociopolitical order.[4] As early as the beginning of the sixteenth century, Niccolò Machiavelli made an influential attempt to redefine political action in terms of instrumental-technical skills rather than religious or moral considerations.[5] Often modeled after Western ideas of causation in the physical world which presupposed "continuity through time and contiguity in space,"[6] causation in the sociopolitical domain appeared to break into the visual field of perception as consisting of a series of directly—and as we shall see also indirectly—observable interactions. In the context of such cultural presuppositions, politics could evolve as a series of events within the temporal and spatial coordinates of the common human experience.

In this framework the use of machines and instruments by political actors, as well as the presentation and justification of action by means of the rhetoric of instrumental or technical rationality, could be seen as part of a general cultural strategy to recast politics in material-visual terms as an inclusive public process. The trend to present the actions of governments in terms of technically conceived causes and effects that are at least partly transparent to the larger public has been constitutive of the modern democratic conception of accountability. Within this conception, the modern democratic state and its agencies could appear as principal technical performers in vast and diverse areas of public action such as defense, health, the economy, transportation, welfare, public architecture, and the like. As such the actions of the governors could appear subject to technical standards of performance and their evaluation by professional or quasi-professional standards could replace, when convenient, more politically problematic assessments that refer to personal, ideological, religious, or ethical norms.

The vast literature which focuses on the question of whether the modern democratic state has succeeded in enlisting science and technology for the enhancement of the instrumental rationality of its actions, or for substantially advancing goals such as security and welfare, tends to ignore or eclipse the enormous *symbolic* functions of technology in the redefinition of liberal-democratic conceptions of power and accountability. Moreover, Continental theorists, generalizing from the particular European experience, have tended to view the links between technology and politics from the more one-sided perspective of the uses of power to control and manipulate while neglecting the role of technology in mediating the accountability of the state to the citizens.[7] In an effort to begin to redress this imbalance I shall confine myself in the following discussion to examining the ways in which the modern liberal democratic state, particularly (although by no means exclusively) the American variant, has appropriated and adapted technology as a political resource for the construction of its particular system of political accountability. I would like to suggest that in developing this system of accountability the liberal-democratic polity had to overcome two dilemmas. The first, inherent in the liberal democratic commitment to the individual as the ultimate unit of action and the ultimate source of values and authority, is whether public order, or more specifically public action, can be generated within an association of private persons. The second dilemma revolves around the need to mitigate the tension between the principle of inclusive political participation and the realities of unevenly distributed political power, the necessity of legitimating leadership within a polity based on a commitment to the values of freedom and self-government and of reconciling anti-power orientations toward the government with the imperatives of governing and exercising power.

I shall suggest that these difficulties have been partly resolved in the modern liberal-democratic polity by the complementary strategies of rationalizing the eye of the democratic citizen as a reliable instrument for establishing and judging "political facts" and of technicalizing public (political) actions as visible, transparent factual events produced by voluntary and therefore accountable agents.[8]

J.-J. Rousseau used the term *civil religion* to indicate the "articles" or "social sentiments without which a man cannot be a good citizen or a faithful subject."[9] Civil religion includes, according to Rousseau, such "positive dogmas" as "the happiness of the just, the punishment of the wicked, the sanctity of the social contract and the laws."[10]

I would like to suggest that, by analogy to Rousseau's concept of civil religion, the liberal democratic polity depends also upon a set of beliefs, or, to use his language, "positive dogmas," concerning the elements that constitute knowledge of political reality, the means to obtain such knowledge, and the persons in whom such knowledge is vested. Together these dogmas constitute what I would describe as the "civil epistemology" of the modern liberal democratic state.

Democratic Political Epistemology and Accountability

Without attempting a detailed description and analysis of the "positive dogmas" which constitute the civil epistemology of the modern democratic polity, I would like to emphasize the centrality of the belief that the exercise of political power can be at least partly transparent to the lay citizens in the field of commonsense experience, the belief that at least in some respects seeing (directly or, as we shall see even more importantly, indirectly) is knowing, that politics, very much like the physical world, is a view, that the visually accessible surfaces of the political world are reliable indicators of factual political truths, that what is seen is largely what is happening, and finally that unlike the celebratory gaze of the credulous subjects gaping at the adorned, opaque, symbolic spectacle of monarchic (or, for that matter, fascist and other hierarchical, authoritarian) power, the gaze of democratic citizens is "attestive," documentary, and critical.

Leading spokespersons and observers of modern democracy such as Thomas Jefferson, Joseph Priestley, Thomas Paine, and Alexis de Tocqueville have stressed the centrality of the transparency of the exercise of political power and the authority of the citizens as spectators-witnesses in upholding democratic accountability and legitimations. Tocqueville is particularly clear on this point when he observes that "equality begets in man the desire of judging of everything for himself, it gives him in all things a taste for the tangible and the real."

> It is on their own testimony [that democratic citizens] are accustomed to rely. They like to discern the object which engages their attention with extreme clearness [and] they therefore strip off as much as possible all that covers it. This disposition of mind soon leads them to condemn forms which they regard as useless or inconvenient veils placed between them and the truth.[11]

The implicit dogmas of democratic civil epistemology are discernible also in Joseph Priestley's insistence that governors show the people that they use their power to "the general advantage" by avoiding "idle pageantry" and concentrating instead on building such things as "noble roads" and "libraries."[12]

The belief that the citizens gaze at the government and that the government makes its actions visible to the citizens is, then, fundamental to the democratic process of government. The shift from the projection of power through pomp and splendor to the projection of power through actions which are either literally technical or at least metaphorically instrumental is, in this context, responsive to the taste Tocqueville ascribes to democratic citizens for "the tangible and the real." The political significance of acting technically in the democratic field of action lies precisely in the supposed anti-theatricality of technology.

While individuals need to learn to discipline their eyes in order to be able to exercise the "attestive gaze" as assertive citizen-witnesses rather than the "celebratory gaze" of docile admiring subjects, the governors need to discipline their actions in order to fit them into the partially visible means—ends schema within which they can be observed and judged as public facts. The technicalization of the actions of the governors can be construed therefore, at least in part, as a strategy of legitimating the exercise of political power by externalizing or objectifying the actions of individual agents and rendering them public. While in non-liberal political contexts the technical objectification of government actions can be carried out precisely in order to elude the pinning down of accountabilities and responsibilities on individual agents, in the liberal-democratic context such tendencies may be partly checked by powerful legal, political, and social restraints and a host of procedures to preserve the category of the agent as the ultimate carrier of responsibility.[13] It is in this connection that, in democracy, transparency and visibility are meant to be employed in order to expose actors to the critical gaze of citizen-witnesses, or their agents, and to check the potential utilization of technicalization as a strategy for escaping politically damaging exposures and attributions of responsibility. When they are welded together and integrated into the liberal-democratic practice of politics, the strategies of technicalizing the governor's actions and of rationalizing the citizen's gaze can balance the empowerment of the governors as actors by the empowerment of the citizens as legitimators or delegitimators of their actions.[14]

While the technicalization of the exercise of political power has served in the liberal-democratic polity as a means to add an objective, public dimension to the subjective elements of action, the regimentation of the citizen's gaze has served to aggregate individual acts of seeing and witnessing as building blocks of a public visual domain. Both strategies construct bridges between the private and the public domains; both permit individuals to function as agents of public actions and public judgments, as producers and spectators of "objective facts," without obliterating the individual subjective components of their status as either voluntary agents who act or as the autonomous self-disciplined spectators who witness and judge. The potential objectification of both action and vision serves here to mediate, rather than eliminate, the separation between the subjective and the objective elements of which acting and seeing in the political context are composed.

I have discussed elsewhere[15] the role of the Western experimental scientific tradition in generating the cultural strategies of technicalizing action and regimenting the eye as a critical-analytical tool for the establishment of facts. The processes through which technical action and visual-witnessing have developed within the scientific tradition and the forces which have led to their conversion into political resources for the redefinition of political power and accountability in the modern democratic state are, of course, too complex to be described or critically examined in this brief discussion. At the deliberate risk of ignoring many important factors, I would like to suggest, however, that there are two principal ways of interpreting these developments. The one takes the view that the technicalization of action and the objectification of sight are important dimensions of the processes of rationalization and modernization in the West, that consistent with the Enlightenment vision of progress such convergences in the cultural foundations of science and politics represent shifts from mythological-religious to more advanced modern cultures.

The other view starts from skeptical and often culturally relativist notions about the nature and the historical development of human knowledge and ends with the view that what is accepted as "real" is socio-historically variable and dependent upon the synthesis of metaphors, conventions, fictions, visual experiences, narratives, and orientations which produce the definitions and the experiences of the real in any given society. Modernization or "progress," according to this view, is largely related to recent transformations in the cultural materials which produce the categories of the real in the last centuries and participate in the literalization and enactment of new images of politics, order, power, and authority.

According to one widely accepted version of the advocates of the former view (I'll call them for convenience "realists"), what guarantees agreements on facts in modern societies are not shared conventions and codes that serve to socialize the multitude into converging productions of the experience, or the effects, of the "real" but rather a given unity of the world as an object of human experience and knowledge which the objectification and the rational regimentation of our sense perceptions allow us to discover. By contrast, representatives of the latter view (the "constructionists") insist that in a state of irremediable ignorance and cultural variability we must constantly shape and reshape our experience and invent notions and categories which are constitutive of our images of nature, society, power, and the like.

While for the realists the Enlightenment is a narrative about the progress of the human species, for the constructionists it is a narrative about changes in the cultural codes with which such basic ideas as the "real," the "just," and the "beautiful" are defined and produced as constituents of historically distinct configurations of order.

The distance between these two views may appear to narrow if we accept the position that while one may not be able to distinguish firmly or absolutely between facts and fictions, neither does one need to depend on such distinctions in order to agree that collectives develop a somewhat shared sense as to which codes, metaphors, conceptions of the real and the fictitious, or which orientations, are better for organizing experience and orienting actions and communications. Societies generate and deploy ideas and views of the world which seem better to fit general, although often only implicitly expressed, preferences concerning the shape of collective life. From this perspective science, technology, and the attestive gaze have been selected out in Western society not because they associate with progress toward objective universalistic truths but because they have appeared more supportive, more constitutive of certain notions of politics, society, and culture which we have come to prefer over others which we have gradually discarded.

Democratic civil epistemology can be described then as a set of commitments and beliefs which enlist such cultural resources as the technological-instrumental forms of action, and what I have called the "attestive visual gaze"[16] to strategies for defining, representing, and holding political power accountable in particularly liberal-democratic ways.

The Production of Democratic Fictions of the Real

The constitutive status of civil epistemology in relation to democratic politics can be clarified further by inquiries into the production of the experience of the "real" and the "natural" in science and painting. Recent historians and sociologists of science have shown the great variety of ways in which the categories of the "real" and the "natural" have been formed and deployed in the experimental, natural scientific tradition.[17] Correspondingly, cultural historians of the visual arts have been presenting increasingly richer accounts of the ways in which painters have been utilizing established cultural codes for defining the "real" and the "natural" and enlisting their supportive optical illusions to produce the experience of reality in spectators of their paintings.[18] In the contexts of both science and art the fictions of the real which work, those which can produce credible experiences of the real and the natural, have been made of complex combinations of actions, concepts, gestures, pictures, linguistic codes, etc.[19] Of particular significance to my concerns here is the important role of machines and instruments in producing "facts" before lay audiences. Robert Boyle, who so skillfully used his air pump to produce factual statements in support of his claims, is a central, even historically paradigmatic, example.[20] As a scientific machine, the air pump had the advantage of appearing to convey truths in the language of sense experience accessible to a wide public lacking in theoretical or mathematical competence. The perceived capacity of machines to convey such messages, to demonstrate facts, without depending on the use of words, had a particular rhetorical power in a society which had learned to suspect the force of words as the coin of wits and rhetoricians.[21]

As Steven Shapin and Simon Schaffer have shown, Boyle devised a host of strategies and techniques for socially augmenting and diffusing the power of the experimental situation to establish facts among publics that did not actually attend and witness the event. They aptly call the methods for certifying the real and the actual to audiences beyond the experimental situation a social technology of "virtual witnessing." "Virtual witnessing" is achieved, they observe, when writers succeed in "triggering in the readers" mind a naturalistic image of the experimental scene."[22] In order to achieve such results, Boyle employed plain, unadorned language, paid much attention to describing the details of the experimental situation, and explicitly or implicitly presented the experimenter as a disinterested observer, sometimes erring or failing to come up with the expected results, a modest man of knowledge distrustful of the use of excessive language and shying away from rhetorical techniques of persuasion.[23] As a rhetorical enterprise, the power of the experimental situation resided, then, in its arrangement and representation as essentially an anti-rhetorical, anti-theatrical, partly risky, mode of persuasion. Experiments, machines, and instruments were designed to produce self-evident facts which "speak for themselves." The power of such facts to "speak for themselves" was connected with the appearance of their independence from makers or operators of the machines and the instruments. It was essential that these tools could fail or produce facts which contradicted the expectations of the experimenters. Boyle was a pioneer of such an expanding practice of experimental persuasion whose impact on Western concepts of the "real," the "natural," and the "artificial" cannot be exaggerated. It is interesting to note how nearly two hundred years later similar subtle and elaborate means were used by a prominent nineteenth-century representative of the experimental scientific tradition, Michael Faraday, to present the facts produced by experiment as unmediated nature revealed directly to the observers undisturbed by human agencies or instruments.[24]

During the nineteenth century the camera as the supposed embodiment of the ideal of mechanized and therefore also objectified production of pictures emerged as an increasingly popular means of virtual witnessing. While the "realists" typically interpret this process as the de-subjectivization, the rationalization of the modern gaze at the world, the "constructionists" draw on the evidence that "photographic viewing" is a cultural construct which developed prior to the invention of the camera and that, as Peter Galassi points out, "photography was not a bastard left by science on the doorstep of art but a legitimate child of the western pictorial tradition."[25] Still, as a machine with automatic parts, the camera appeared to open the way for substituting reliance on potentially untrustworthy observers by reliance on mechanical devices.[26] Although artist-photographers such as Alfred Stieglitz observed and demonstrated in the early twentieth century the uses of the camera like the painter's brush, as a tool for the production of authentic subjective visions, in the field of journalism, news, and politics the camera emerged as a means to secure neutral optical representations of the "real world." Again, then, the "real" came to be connected with mechanical production of facts, in this case "pictures." The popularization of the camera could transform any human subject into a potential producer of neutral-documentary representations of the "real" and a source of photographs, that is, resources of virtual witnessing, for one's fellow men and women. This development (as the cultural constructionists would argue) cannot be attributed simply to the technological inventions which led to the camera. It appears to be even more strongly related to cultural and political processes which, consistent with the evolving dogmas of democratic civil epistemology, standardized and disciplined the attestive gaze as the shared visual strategy of a multitude of individual observers. Jonathan Crary seems to support this view in his instructive, detailed discussion of the role of the camera obscura and then of the camera as "techniques of the observer" and their dependence upon prior ideological and cultural commitments.[27] Crary insists that:

> Central components of nineteenth century realism, of mass visual culture, preceded the invention of photography and in no way required photographic procedures or even the development of mass production techniques. Rather they are inextricably dependent on a new arrangement of knowledge about the body and the constitutive relation of that knowledge to social power. These apparatuses are the outcome of a complex remaking of the individual as observer into something calculable and regularizable, and of human vision into something measurable and thus exchangeable. The standardization of visual imagery in the nineteenth century must be seen then not simply as part of new forms of mechanized reproduction but in relation to a broader process of normalization and subjection of the observer.[28]

An important source of the force of the photograph as a visual resource of virtual witnessing was the fact that unlike painting, photographing, as a much more instant act, appeared less amenable to control by a long careful application of the photographer's intention and the camera could be used in ways which would limit the capacity of the subject photographed to stage his or her appearance, thus producing more candid photographs of human beings. Norman Bryson is instructive on this point when he suggests that in our culture the illusion of the real depends on the apparent absence of a manifestly controlling intention, on the appearance of the contingent, chancy, and at least in part the disorderly.[29] Hence the experience of the "real" is achieved espe-cially when the contingent, captured by mechanical visual recording, is juxtaposed with the experience of the staged, or the theatrically contrived. It is as if in order to be persuasive in producing the effects of the real, literalizers as the producers of self-de-nying fictions of the real need the fictionalizers as the producers of self-proclaiming fictions.

The power of sociopsychological processes to enlist the sense of sight to produce shared commonsense notions of the real has often been noticed by scientists who have experienced the tensions between social and specialized scientific ideas or construc-tions of reality. Particularly instructive in this connection are Einstein's observations, in his discussion of the experience of lightning, on how "false" commonsense notions are formed. He suggests that at first the experience "it is lightning" appears as a per-sonal experience. But

> when the person who has had the experience of lightning also *experiences other persons in a way that brings their behavior into relation with his own experience,* "it is lightning," that experience is no longer interpreted as an exclusively personal experience but as an experience of other persons. . . . Thus what started as a private experience becomes interpreted as an objective event.[30] (my emphasis)

The subjective visual experience combined with the experience of the behavior of other persons becomes translated into notions of real facts.

Against the background of such accounts, it should perhaps be less difficult to recognize the relevance of the "positive dogmas" of democratic civil epistemology to the sanctioning of certain slices of experience, certain gestures, visual experiences, utterances, and fictions as the raw materials of political reality, of facts certified for social and political currency.

An important feature of liberal-democratic constructions of the facts of actions and events is the omnipresence of the contingent, the unpredictable, and the uncon-trollable. Just as Boyle's authority as an honest and trustworthy experimenter de-pended, among other things, upon his readiness to expose his failures and errors as well as his successes, so democratic governments cannot avoid, if they wish to gain trust, the political risks of transparent failures or of contingencies which publicly defy well-entrenched expectations. It is in manifestly defying such expectations that the contingent and the unpredictable validate a dimension of our experience which ap-pears to resist our anticipations, and often our wishes, as "reality," as something exter-nal and partly independent of us. In totalitarian societies such as the Soviet Union, observed Elihu Katz and Daniel Dayan (in 1985), "one has no chance to see the live broadcast of a Russian space missile lest it explode on takeoff. . . . Funerals and anni-versaries are safer."[31] They are safer because they can be more theatrically controlled. In democracy, technology is integrated by contrast into a political spectacle of actions and events which are at least partly but unavoidably unpredictable and resistant to staging. It is precisely the inherent presence of the contingent and the uncertain that makes technology appear more "real," more authentic than such staged spectacles as state funerals. Action in the democratic public arena is therefore a visibly more evenly balanced interaction between agency and situation than in the monarchy or the fascist state where agency is supposedly much more in control.

Inherent uncertainties and contingencies, then, humble and level democratic actors. But they make the democratic spectacle a less boring and more attractive synthesis between the predictable and the surprising. Precisely because in some respects "facts" which describe political actions and events in the liberal democracy are less determined and fixed than "facts" in a totalitarian state, they can function as substitutes for alternative, more authoritarian closures in the context of political discourses and actions. It is as if such "facts" had become flexibly fixed as neutral or external standards for tempering conflicts of opinion by more softly adjudicating, or draining, opposition among the various actors. Dogmas such as the claim that democratic governments are transparent to their citizens or that technology partly objectifies action and therefore furnishes at least a partial escape from the inherent indeterminism of the ethical and the political have come, then, to serve the specific requirements of a particular conception of order.

Modern mass-printed and electronic media, employing in part variants of the techniques of virtual witnessing devised and used by experimental scientists, have become a principal means for the production and diffusion of certified fictions of the politically real. While from within the ethos of journalism, as one of the principal expressions of the dogmas of democratic civil epistemology, the role of the press is defined as neutrally mirroring social and political reality, subtle analysts and historians have alluded to the specific social conventions and cultural codes which are enlisted in the very creation of the shared experiences which come to stand for sociopolitical realities.[32] As one of these codes, modern democratic civil epistemology does not lead us to discover a given world of social or political facts but is rather itself a powerful device for enacting politics as a view, political actors as performers, journalists as observers (who translate actual into virtual witnessing), and citizens as witnesses.

Not surprisingly, the power of cultural strategies grounded in democratic civil epistemology and technology to create the experiences of social and political reality has affected wide spheres of social and cultural activity beyond the political domain. A most instructive example is the penetration in the course of the eighteenth and nineteenth centuries of the detached, unadorned language of "reportage" into literature where it was associated with the rise of the realistic novel as a new literary genre,[33] and to historiography where it became associated with the redefinition of canons of history writing.[34]

Gradually techniques of virtual witnessing have come to be more consciously employed not so much to represent and transmit authentic experiences to others as to actually employ particular conventions for producing the effects of the real by mixing elements of experience, of the fictitious, and of the contrived.

The social entrenchment of the pictorial techniques of producing in spectators a sense of literally contacting, or witnessing, the "real" is indicated in the force of works of art which alert or provoke the spectator to recognize reflexively how his illusions of the real are formed. Magritte's painting is a case in point. His paintings often impose on their spectators the uneasy yet liberating discovery of the fictitious basis of the optical surface-experience of the real. It is precisely this ability to subvert in people the illusion that certain fictions stand for real things which made Magritte so interesting to Foucault and Derrida as critics of the referential theories of language and painting.[35]

It is also because of the fragility of the fictitious experience of the real, or more precisely of the democratic impulse to pin down responsibilities for the exercise of political power in the elusive universe of human actions by fixing specific fictions of the "real," that the instrumental technicalization of action has been so useful a political strategy in the modern democratic polity. It is largely because of the claim or belief that technology physicalizes and objectifies action, that it can translate action into actual or imagined material causes and effects, into events, that technology could appear to respond to the democratic need to confine and fix accountabilities within particular coordinates of time and space.

As such, technology experienced and interpreted within a democratic civil epistemology has produced and upheld a modern democratic concept of visible power whose exercise appears publicly accountable to the large public.

The welding of technology, political action, and accountability within the framework of the "positive dogmas" of democratic civil epistemology has constituted a powerful response to the two principal dilemmas of liberal-democratic politics I discussed above: the problem of establishing private persons as the agencies of public actions and the problem of legitimating the uneven exercise of the power to act in a polity committed to self-government and distrustful of strong leadership. The technicalization of action in the political field has emerged as a strategy partly to depersonalize action and detach it from the acting persons and partly to objectify it as an object of the attestive-visual gaze. Thus the visibility of action in a field of partly contingent facts and events has become a means to check, humble, and ultimately delegitimate and replace democratic leaders as performers. The universalization of accountability through exposure in a public visual field evolved as a principal strategy for (always temporarily) legitimating the actual centralization of power.

In the late twentieth century, however, the increasingly pronounced distrust of the possibility of fixed neutral points of view, the resubjectivization of the "attestive individual gaze," and the increasingly acknowledged contributions of observers to "making" what they see, and the repoliticization as well as the remoralization of technological action indicate the disintegration of the Enlightenment synthesis which underlay the modern enactment of democratic politics.[36] Postmodernism in this context is largely the spread of reflexive orientations which acknowledge the self-denying theatricality of democratic politics as an exchange of fictions of the politically real between virtually self-exposing political performers and their virtual critical witnesses.

Notes and References

1. Yaron Ezrahi, *The Descent of Icarus: Science and the Transformation of Contemporary Democracy* (Cambridge, MA: Harvard University Press, 1990).

2. Thomas Paine, *The Rights of Man*. See the citation and the discussion by W. J. T. Mitchell, *Iconology, Image, Text, Ideology* (Chicago: University of Chicago Press, 1986), pp. 141–42.

3. Thomas Paine, *The Rights of Man: An Answer to Mr. Burke's Attack on the French Revolution* (London: Carlite, 1819), p. 36.

4. Don Handelman, *Models and Mirrors: Towards an Anthropology of Public Events* (Cambridge: Cambridge University Press, 1990). See also Sheldon Wolin, "Postmodern Politics and the Absence of Myth," *Social Research* 52 (1985), pp. 217–39 (cited in Handelman).

5. Niccolò Machiavelli, *The Prince and the Discourses* (New York: The Modern Library, 1950).

6. Handelman, *Models and Mirrors*, op. cit., p. 29.

7. See, for instance, Jacques Ellul, *The Technological Society* (New York: Vintage Books, 1964); also Michel Foucault's *Discipline and Punish* (New York: Pantheon Books, 1977) and Herbert Marcuse, *One-Dimensional Man* (Boston: Beacon Press, 1964).

8. For a fuller discussion see my *The Descent of Icarus*, op. cit.

9. Jean-Jacques Rousseau, *The Social Contract and Discourses*, G. D. H. Cole (trans.) (New York: E. P. Dutton and Co., 1950), p. 139.

10. Ibid.

11. Alexis de Tocqueville, *Democracy in America*, vol. II. Phillips Bradley (ed.) (New York: Vintage Books, 1945), vol. II, pp. 4–5, 42.

12. Joseph Priestley, *Priestley's Writings on Philosophy, Science and Politics*, ed. and intro. J. A. Passmore (New York: Collier, 1965), p. 254.

13. The inquiry into the responsibility of individual decision makers working within the American space agency in connection with the explosion of Challenger-2 is a characteristic illustration.

14. See in my *The Descent of Icarus*, op. cit.

15. Ibid.

16. Ibid., pp. 69–127.

17. See, e.g., Steven Shapin and Simon Schaffer, *Leviathan and the Air Pump, Hobbes, Boyle and the Experimental Life* (Princeton: Princeton University Press, 1985); Bruno Latour and Steve Woolgar, *Laboratory Life: The Construction of Scientific Facts* (Princeton: Princeton University Press, 1986), and David Gooding, Trevor Pinch, and Simon Schaffer (eds.), *The Uses of Experiment* (Cambridge: Cambridge University Press, 1989).

18. See, e.g., Norman Bryson, *Vision and Painting, The Logic of the Gaze*; Svetlana Alpers, *The Art of Describing* (Chicago: University of Chicago Press, 1983); Michael Fried, *Absorption and Theatricality* (Berkeley: University of California Press, 1980).

19. See references 16 and 17 above.

20. Shapin and Schaffer, *Leviathan and the Air Pump*, op. cit.

21. See one of the earliest articulations of this position in Thomas Sprat, *The History of the Royal Society of London for the Improving of Natural Knowledge* (London: 1667).

22. Shapin and Schaffer, *Leviathan and the Air Pump*, op. cit., pp. 60–65, 225–26, 327, 336.

23. Ibid.

24. D. C. Gooding and F. A. J. L. James (eds.), *Faraday Rediscovered* (London: Macmillan/New York: Stockton Press, 1985).

25. Peter Galassi, *Before Photography* (New York: Museum of Modern Art, 1981), p. 12.

26. Stanley Cavell, *The World Viewed*, enlarged ed. (Cambridge, MA: Harvard University Press, 1979).

27. Jonathan Crary, *Techniques of the Observer: On Vision and Modernity in the Nineteenth Century* (Cambridge, MA: MIT Press, 1990).

28. Ibid.

29. Bryson, *Vision and Painting, The Logic of the Gaze*, op. cit., pp. 64–65.

30. See a more detailed discussion in Yaron Ezrahi, "Einstein and the Light of Reason" in Gerald Holton and Yehuda Elkana (eds.), *Albert Einstein: Historical and Cultural Perspectives* (Princeton: Princeton University Press, 1982), pp. 256–58.

31. Elihu Katz and Daniel Dayan, "Media Events: On the Experience of Not Being There," *Religion* 15 (1985): 306.

32. See, for instance, Gaye Tuchman, *Making News: A Study in the Construction of Reality* (New York: Free Press, 1978); Michael Schudson, *Discovering the News* (New York: Basic Books, 1978).

33. Leonard J. Davis, *Factual Fictions: The Origins of the English Novel* (New York: Columbia University Press, 1983); George J. Becker (ed.), *Documents of Modern Literary Realism* (Princeton: Princeton University Press, 1967).

34. Peter Novick, *That Noble Dream, The "Objectivity Question" and the American Historical Profession* (Cambridge: Cambridge University Press, 1988).

35. Michel Foucault, *This is Not a Pipe*, ed. and trans. James Harkness (Berkeley: University of California Press, 1982); Jacques Derrida, *The Truth in Painting*, trans. G. Bennington and Ian McLeod (Chicago: University of Chicago Press, 1987).

36. *The Descent of Icarus*, op. cit., part III.

PART V

Feminist Perspectives: Knowledge and Bodies

THE TOTALIZATION OF technique incorporates the human body into the system. No longer the locus of the natural and spontaneous, it becomes the site of multifarious technological interventions. At the same time, individuals are socially situated through the significance attached to their bodily existence, through gender, race, and nationality. This section offers two approaches to the body, as the basis of finite knowledge, and as a (re)productive resource.

Donna Haraway's "Situated Knowledges: The Science Question in Feminism and the Privilege of Partial Perspective" lies at the intersection of feminism and constructivism and adds something new to both. Haraway insists on the essentially local character of every form of knowledge, by which she means its inevitable rootedness in a body and a social situation. Yet at the same time she resists the relativistic conclusion that so often derives from this premise. If the very idea of an absolute standpoint is meaningless, then no skeptical conclusion need follow from the discovery that human beings cannot occupy it. Instead, we need a conception of finite knowledge as an interactive process confronting the fruits of the various socially relative standpoints that are humanly available.

In "Knowledge, Bodies, and Values: Reproductive Technologies and Their Scientific Context," Helen E. Longino offers a post-empiricist critique of current reproductive biology and its medical applications. Her argument resumes constructivist themes developed by Feenberg, Vogel, Ezrahi, and Haraway. Because scientific theories are underdetermined by observable facts, evaluative presuppositions must intervene in deciding between plausible alternative accounts of the evidence. Feminist analysis brings to light the role of masculinist visions in the shaping of reproductive science and technology. The struggle for political enlightenment can play a corresponding role in freeing science and technology of the burden of masculinist bias and thus open new paths for research.

11

Situated Knowledges

The Science Question in Feminism and the Privilege of Partial Perspective

Donna Haraway

A CADEMIC AND ACTIVIST feminist inquiry has repeatedly tried to come to terms with the question of what *we* might mean by the curious and inescapable term *objectivity*. We have used a lot of toxic ink and trees processed into paper decrying what *they* have meant and how it hurts *us*. The imagined "they" constitute a kind of invisible conspiracy of masculinist scientists and philosophers replete with grants and laboratories. The imagined "we" are the embodied others, who are not allowed *not* to have a body, a finite point of view, and so an inevitably disqualifying and polluting bias in any discussion of consequence outside our own little circles, where a "mass"-subscription journal might reach a few thousand readers composed mostly of science haters. At least, I confess to these paranoid fantasies and academic resentments lurking underneath some convoluted reflections in print under my name in the feminist literature in the history and philosophy of science. We, the feminists in the debates about science and technology, are the Reagan era's "special-interest groups" in the rarified realm of epistemology, where traditionally what can count as knowledge is policed by philosophers codifying cognitive canon law. Of course, a special-interest group is, by Reaganoid definition, any collective historical subject that dares to resist the stripped-down atomism of Star Wars, hypermarket, postmodern, media-simulated citizenship. Max Headroom doesn't have a body; therefore, he alone *sees* everything in the great communicator's empire of the Global Network. No wonder Max gets to have a naive sense of humor and a kind of happily regressive, preoedipal sexuality, a sexuality that we ambivalently—with dangerous incorrectness—had imagined to be reserved for lifelong inmates of female and colonized bodies and maybe also white male computer hackers in solitary electronic confinement.

It has seemed to me that feminists have both selectively and flexibly used and been trapped by two poles of a tempting dichotomy on the question of objectivity. Certainly I speak for myself here, and I offer the speculation that there is a collective discourse on these matters. Recent social studies of science and technology, for example, have made available a very strong social constructionist argument for *all* forms of knowledge claims, most certainly and especially scientific ones.[1] According to these tempting views, no insider's perspective is privileged, because all drawings of inside-outside boundaries in knowledge are theorized as power moves, not moves toward truth. So, from the strong social constructionist perspective, why should we be cowed by scientists' descriptions of their activity and accomplishments; they and their patrons have stakes in throwing sand in our eyes. They tell parables about objectivity and scientific method to students in the first years of their initiation, but no practitioner of the high scientific arts would be caught dead *acting on* the textbook versions. Social constructionists make clear that official ideologies about objectivity and scientific method are particularly bad guides to how scientific knowledge is actually *made*. Just as for the rest of us, what scientists believe or say they do and what they really do have a very loose fit.

The only people who end up actually *believing* and, goddess forbid, acting on the ideological doctrines of disembodied scientific objectivity—enshrined in elementary textbooks and technoscience booster literature—are nonscientists, including a few very trusting philosophers. Of course, my designation of this last group is probably just a reflection of a residual disciplinary chauvinism acquired from identifying with historians of science and from spending too much time with a microscope in early adulthood in a kind of disciplinary preoedipal and modernist poetic moment when cells seemed to be cells and organisms, organisms. *Pace*, Gertrude Stein. But then came the law of the father and its resolution of the problem of objectivity, a problem solved by always already absent referents, deferred signifieds, split subjects, and the endless play of signifiers. Who wouldn't grow up warped? Gender, race, the world itself—all seem the effects of warp speeds in the play of signifiers in a cosmic force field.

In any case, social constructionists might maintain that the ideological doctrine of scientific method and all the philosophical verbiage about epistemology were cooked up to distract our attention from getting to know the world *effectively* by practicing the sciences. From this point of view, science—the real game in town—is rhetoric, a series of efforts to persuade relevant social actors that one's manufactured knowledge is a route to a desired form of very objective power. Such persuasions must take account of the structure of facts and artifacts, as well as of language-mediated actors in the knowledge game. Here, artifacts and facts are parts of the powerful art of rhetoric. Practice is persuasion, and the focus is very much on practice. All knowledge is a condensed node in an agonistic power field. The strong program in the sociology of knowledge joins with the lovely and nasty tools of semiology and deconstruction to insist on the rhetorical nature of truth, including scientific truth. History is a story Western culture buffs tell each other; science is a contestable text and a power field; the content is the form.[2] Period.

So much for those of us who would still like to talk about *reality* with more confidence than we allow to the Christian Right when they discuss the Second Coming and their being raptured out of the final destruction of the world. We would like to think our appeals to real worlds are more than a desperate lurch away from cynicism and an act of faith like any other cult's, no matter how much space we generously give to all the rich and always historically specific mediations through which we and everybody else must know the world. But the further I get in describing the radical social constructionist program and a particular version of postmodernism, coupled with the acid tools of critical discourse in the human sciences, the more nervous I get. The imagery of force fields, of moves in a fully textualized and coded world, which is the working metaphor in many arguments about socially negotiated reality for the postmodern subject, is, just for starters, an imagery of high-tech military fields, of automated academic battlefields, where blips of light called players disintegrate (what a metaphor!) each other in order to stay in the knowledge and power game. Technoscience and science fiction collapse into the sun of their radiant (ir)reality—war.[3] It shouldn't take decades of feminist theory to sense the enemy here. Nancy Hartsock got all this crystal clear in her concept of abstract masculinity.[4]

I, and others, started out wanting a strong tool for deconstructing the truth claims of hostile science by showing the radical historical specificity, and so contestability, of *every* layer of the onion of scientific and technological constructions, and we end up with a kind of epistemological electroshock therapy, which, far from ushering us into the high-stakes tables of the game of contesting public truths, lays us out on the table with self-induced multiple-personality disorder. We wanted a way to go beyond showing bias in science (that proved too easy anyhow) and beyond separating the good scientific sheep from the bad goats of bias and misuse. It seemed promising to do this by the strongest possible constructionist argument that left no cracks for reducing the issues to bias versus objectivity, use versus misuse, science versus pseudo-science. We unmasked the doctrines of objectivity because they threatened our budding sense of collective historical subjectivity and agency and our "embodied" accounts of the truth, and we ended up with one more excuse for not learning any post-Newtonian physics and one more reason to drop the old feminist self-help practices of repairing our own cars. They're just texts anyway, so let the boys have them back.

Some of us tried to stay sane in these disassembled and dissembling times by holding out for a feminist version of objectivity. Here, motivated by many of the same political desires, is the other seductive end of the objectivity problem. Humanistic Marxism was polluted at the source by its structuring theory about the domination of nature in the self-construction of man and by its closely related impotence in relation to historicizing anything women did that didn't qualify for a wage. But Marxism was still a promising resource as a kind of epistemological feminist mental hygiene that sought our own doctrines of objective vision. Marxist starting points offered a way to get to our own versions of standpoint theories, insistent embodiment, a rich tradition of critiquing hegemony without disempowering positivisms and relativisms and a way to get to nuanced theories of mediation. Some versions of psychoanalysis were of aid in this approach, especially anglo-phone object relations theory, which maybe did more for U.S. socialist feminism for a time than anything from the pen of Marx or Engels, much less Althusser or any of the late pretenders to son-ship treating the subject of ideology and science.[5]

Another approach, "feminist empiricism," also converges with feminist uses of Marxian resources to get a theory of science which continues to insist on legitimate meanings of objectivity and which remains leery of a radical constructivism conjugated with semiology and narratology.[6] Feminists have to insist on a better account of the world; it is not enough to show radical historical contingency and modes of construction for everything. Here, we, as feminists, find ourselves perversely conjoined with the discourse of many practicing scientists, who, when all is said and done, mostly believe they are describing and discovering things *by means of* all their constructing and arguing. Evelyn Fox Keller has been particularly insistent on this fundamental matter, and Sandra Harding calls the goal of these approaches a "successor science." Feminists have stakes in a successor science project that offers a more adequate, richer, better account of a world, in order to live in it well and in critical, reflexive relation to our own as well as others' practices of domination and the unequal parts of privilege and oppression that make up all positions. In traditional philosophical categories, the issue is ethics and politics perhaps more than epistemology.

So, I think my problem, and "our" problem, is how to have *simultaneously* an account of radical historical contingency for all knowledge claims and knowing subjects, a critical practice for recognizing our own "semiotic technologies" for making meanings, *and* a no-nonsense commitment to faithful accounts of a "real" world, one that can be partially shared and that is friendly to earthwide projects of finite freedom, adequate material abundance, modest meaning in suffering, and limited happiness. Harding calls this necessary multiple desire a need for a successor science project and a postmodern insistence on irreducible difference and radical multiplicity of local knowledges. *All* components of the desire are paradoxical and dangerous, and their combination is both contradictory and necessary. Feminists don't need a doctrine of objectivity that promises transcendence, a story that loses track of its mediations just where someone might be held responsible for something, and unlimited instrumental power. We don't want a theory of innocent powers to represent the world, where language and bodies both fall into the bliss of organic symbiosis. We also don't want to theorize the world, much less act within it, in terms of Global Systems, but we do need an earthwide network of connections, including the ability partially to translate knowledges among very different—and power-differentiated—communities. We need the power of modern critical theories of how meanings and bodies get made, not in order to deny meanings and bodies, but in order to build meanings and bodies that have a chance for life.

Natural, social, and human sciences have always been implicated in hopes like these. Science has been about a search for translation, convertibility, mobility of meanings, and universality—which I call reductionism only when one language (guess whose?) must be enforced as the standard for all the translations and conversions. What money does in the exchange orders of capitalism, reductionism does in the powerful mental orders of global sciences. There is, finally, only one equation. That is the deadly fantasy that feminists and others have identified in some versions of objectivity, those in the service of hierarchical and positivist orderings of what can count as knowledge. That is one of the reasons the debates about objectivity matter, metaphorically and otherwise. Immortality and omnipotence are not our goals. But we could use some enforceable, reliable accounts of things not reducible to power moves and agonistic, high-status games of rhetoric or to scientistic, positivist arrogance. This point applies whether we are talking about genes, social classes, elementary particles, genders, races, or texts; the point applies to the exact, natural, social, and human science, despite the slippery ambiguities of the words "objectivity" and "science" as we slide around the discursive terrain. In our efforts to climb the greased pole leading to a usable doctrine of objectivity, I and most other feminists in the objectivity debates have alternatively, or even simultaneously, held on to both ends of the dichotomy, a dichotomy which Harding describes in terms of successor science projects versus postmodernist accounts of difference and which I have sketched in this essay as radical constructivism versus feminist critical empiricism. It is, of course, hard to climb when you are holding on to both ends of a pole, simultaneously or alternatively. It is, therefore, time to switch metaphors.

The Persistence of Vision

I would like to proceed by placing metaphorical reliance on a much maligned sensory system in feminist discourse: vision.[7] Vision can be good for avoiding binary oppositions. I would like to insist on the embodied nature of all vision and so reclaim the sensory system that has been used to signify a leap out of the marked body and into a conquering gaze from nowhere. This is the gaze that mythically inscribes all the marked bodies, that makes the unmarked category claim the power to see and not be seen, to represent while escaping representation. This gaze signifies the unmarked positions of Man and White, one of the many nasty tones of the word "objectivity" to feminist ears in scientific and technological, late-industrial, militarized, racist, and male-dominant societies, that is, here, in the belly of the monster, in the United States in the late 1980s. I would like a doctrine of embodied objectivity that accommodates paradoxical and critical feminist science projects: Feminist objectivity means quite simply *situated knowledges*.

The eyes have been used to signify a perverse capacity—honed to perfection in the history of science tied to militarism, capitalism, colonialism, and male supremacy—to distance the knowing subject from everybody and everything in the interests of unfettered power. The instruments of visualization in multinationalist, postmodernist culture have compounded these meanings of disembodiment. The visualizing technologies are without apparent limit. The eye of any ordinary primate like us can be endlessly enhanced by sonography systems, magnetic resonance imaging, artificial intelligence–linked graphic manipulation systems, scanning electron microscopes, computer tomography scanners, color-enhancement techniques, satellite surveillance systems, home and office video display terminals, cameras for every purpose from filming the mucous membrane lining the gut cavity of a marine worm living in the vent gases on a fault between continental plates to mapping a planetary hemisphere elsewhere in the solar system. Vision in this technological feast becomes unregulated gluttony; all seems not just mythically about the god trick of seeing everything from nowhere, but to have put the myth into ordinary practice. And like the god trick, this eye fucks the world to make techno-monsters. Zoe Sofoulis calls this the cannibaleye of masculinist extra-terrestrial projects for excremental second-birthing.

A tribute to this ideology of direct, devouring, generative, and unrestricted vision, whose technological mediations are simultaneously celebrated and presented as utterly transparent, can be found in the volume celebrating the 100th anniversary of the National Geographic Society. The volume closes its survey of the magazine's quest literature, effected through its amazing photography, with two juxtaposed chapters. The first is on "Space," introduced by the epigraph "The choice is the universe—or nothing."[8] This chapter recounts the exploits of the space race and displays the color-enhanced "snapshots" of the outer planets reassembled from digitalized signals transmitted across vast space to let the viewer "experience" the moment of discovery in immediate vision of the "object."[9] These fabulous objects come to us simultaneously as indubitable recordings of what is simply there and as heroic feats of technoscientific production. The next chapter is the twin of outer space: "Inner Space," introduced by the epigraph "The stuff of stars has come alive."[10] Here, the reader is brought into the realm of the infinitesimal, objectified by means of radiation outside the wave lengths that are "normally" perceived by hominid primates, that is, the beams of lasers and scanning electron microscopes, whose signals are processed into the wonderful full-color snapshots of defending T cells and invading viruses.

But, of course, that view of infinite vision is an illusion, a god trick. I would like to suggest how our insisting metaphorically on the particularity and embodiment of all vision (although not necessarily organic embodiment and including technological mediation), and not giving in to the tempting myths of vision as a route to disembodiment and second-birthing, allows us to construct a usable, but not an innocent, doctrine of objectivity. I want a feminist writing of the body that metaphorically emphasizes vision again, because we need to reclaim that sense to find our way through all the visualizing tricks and powers of modern sciences and technologies that have transformed the objectivity debates. We need to learn in our bodies, endowed with primate

color and stereoscopic vision, how to attach the objective to our theoretical and political scanners in order to name where we are and are not, in dimensions of mental
and physical space we hardly know how to name. So, not so perversely, objectivity
turns out to be about particular and specific embodiment and definitely not about the
false vision promising transcendence of all limits and responsibility. The moral is simple: only partial perspective promises objective vision. All Western cultural narratives
about objectivity are allegories of the ideologies governing the relations of what we
call mind and body, distance and responsibility. Feminist objectivity is about limited
location and situated knowledge, not about transcendence and splitting of subject and
object. It allows us to become answerable for what we learn how to see.

These are lessons that I learned in part walking with my dogs and wondering how
the world looks without a fovea and very few retinal cells for color vision but with a
huge neural processing and sensory area for smells. It is a lesson available from photographs of how the world looks to the compound eyes of an insect or even from the
camera eye of a spy satellite or the digitally transmitted signals of space probe-per-
ceived differences "near" Jupiter that have been transformed into coffee table color
photographs. The "eyes" made available in modern technological sciences shatter any
idea of passive vision; these prosthetic devices show us that all eyes, including our own
organic ones, are active perceptual systems, building on translations and specific *ways*
of seeing, that is, ways of life. There is no unmediated photograph or passive camera
obscura in scientific accounts of bodies and machines; there are only highly specific
visual possibilities, each with a wonderfully detailed, active, partial way of organizing
worlds. All these pictures of the world should not be allegories of infinite mobility and
interchangeability but of elaborate specificity and difference and the loving care people might take to learn how to see faithfully from another's point of view, even when
the other is our own machine. That's not alienating distance; that's a *possible* allegory
for feminist versions of objectivity. Understanding how these visual systems work, technically, socially, and psychically, ought to be a way of embodying feminist objectivity.

Many currents in feminism attempt to theorize grounds for trusting especially the
vantage points of the subjugated; there is good reason to believe vision is better from
below the brilliant space platforms of the powerful.[11] Building on that suspicion, this
essay is an argument for situated and embodied knowledges and an argument against
various forms of unlocatable, and so irresponsible, knowledge claims. Irresponsible
means unable to be called into account. There is a premium on establishing the capacity to see from the peripheries and the depths. But here there also lies a serious
danger of romanticizing and/or appropriating the vision of the less powerful while
claiming to see from their positions. To see from below is neither easily learned nor
unproblematic, even if "we" "naturally" inhabit the great underground terrain of subjugated knowledges. The positionings of the subjugated are not exempt from critical
reexamination, decoding, deconstruction, and interpretation; that is, from both semio-

logical and hermeneutic modes of critical inquiry. The standpoints of the subjugated are not "innocent" positions. On the contrary, they are preferred because in principle they are least likely to allow denial of the critical and interpretive core of all knowledge. They are knowledgeable of modes of denial through repression, forgetting, and disappearing acts—ways of being nowhere while claiming to see comprehensively. The subjugated have a decent chance to be on to the god trick and all its dazzling— and, therefore, blinding—illuminations. "Subjugated" standpoints are preferred because they seem to promise more adequate, sustained, objective, transforming accounts of the world. But *how* to see from below is a problem requiring at least as much skill with bodies and language, with the mediations of vision, as the "highest" technoscientific visualizations.

Such preferred positioning is as hostile to various forms of relativism as to the most explicitly totalizing versions of claims to scientific authority. But the alternative to relativism is not totalization and single vision, which is always finally the unmarked category whose power depends on systematic narrowing and obscuring. The alternative to relativism is partial, locatable, critical knowledges sustaining the possibility of webs of connections called solidarity in politics and shared conversations in epistemology. Relativism is a way of being nowhere while claiming to be everywhere equally. The "equality" of positioning is a denial of responsibility and critical inquiry. Relativism is the perfect mirror twin of totalization in the ideologies of objectivity; both deny the stakes in location, embodiment, and partial perspective; both make it impossible to see well. Relativism and totalization are both "god tricks" promising vision from everywhere and nowhere equally and fully, common myths in rhetorics surrounding Science. But it is precisely in the politics and epistemology of partial perspectives that the possibility of sustained, rational, objective inquiry rests.

So, with many other feminists, I want to argue for a doctrine and practice of objectivity that privileges contestation, deconstruction, passionate construction, webbed connections, and hope for transformation of systems of knowledge and ways of seeing. But not just any partial perspective will do; we must be hostile to easy relativisms and holisms built out of summing and subsuming parts. "Passionate detachment"[12] requires more than acknowledged and self-critical partiality. We are also bound to seek perspective from those points of view, which can never be known in advance, that promise something quite extraordinary, that is, knowledge potent for constructing worlds less organized by axes of domination. From such a viewpoint, the unmarked category would *really* disappear—quite a difference from simply repeating a disappearing act. The imaginary and the rational—the visionary and objective vision— hover close together. I think Harding's plea for a successor science and for postmodern sensibilities must be read as an argument for the idea that the fantastic element of hope for transformative knowledge and the severe check and stimulus of sustained critical inquiry are jointly the ground of any believable claim to objectivity or rationality not riddled with breathtaking denials and repressions. It is even possible to read the record of scientific revolutions in terms of this feminist doctrine of rationality and objectivity. Science has been utopian and visionary from the start; that is one reason "we" need it.

A commitment to mobile positioning and to passionate detachment is dependent on the impossibility of entertaining innocent "identity" politics and epistemologies as strategies for seeing from the standpoints of the subjugated in order to see well. One cannot "be" either a cell or molecule—or a woman, colonized person, laborer, and so on—if one intends to see and see from these positions critically. "Being" is much more problematic and contingent. Also, one cannot relocate in any possible vantage point without being accountable for that movement. Vision is *always* a question of the power to see—and perhaps of the violence implicit in our visualizing practices. With whose blood were my eyes crafted? These points also apply to testimony from the position of "oneself." We are not immediately present to ourselves. Self-knowledge requires a semiotic-material technology to link meanings and bodies. Self-identity is a bad visual system. Fusion is a bad strategy of positioning. The boys in the human sciences have called this doubt about self-presence the "death of the subject" defined as a single ordering point of will and consciousness. That judgment seems bizarre to me. I prefer to call this doubt the opening of nonisomorphic subjects, agents, and territories of stories unimaginable from the vantage point of the cyclopean, self-satiated eye of the master subject. The Western eye has fundamentally been a wandering eye, a traveling lens. These peregrinations have often been violent and insistent on having mirrors for a conquering self—but not always. Western feminists also *inherit* some skill in learning to participate in revisualizing worlds turned upside down in earth-transforming challenges to the views of the masters. All is not to be done from scratch.

The split and contradictory self is the one who can interrogate positionings and be accountable, the one who can construct and join rational conversations and fantastic imaginings that change history.[13] Splitting, not being, is the privileged image for feminist epistemologies of scientific knowledge. "Splitting" in this context should be about heterogeneous multiplicities that are simultaneously salient and incapable of being squashed into isomorphic slots or cumulative lists. This geometry pertains within and among subjects. Subjectivity is multidimensional; so, therefore, is vision. The knowing self is partial in all its guises, never finished, whole, simply there and original; it is always constructed and stitched together imperfectly, and *therefore* able to join with another, to see together without claiming to be another. Here is the promise of objectivity: a scientific knower seeks the subject position, not of identity, but of objectivity, that is, partial connection. There is no way to "be" simultaneously in all, or wholly in any, of the privileged (i.e., subjugated) positions structured by gender, race, nation, and class. And that is a short list of critical positions. The search for such a "full" and total position is the search for the fetishized perfect subject of oppositional history, sometimes appearing in feminist theory as the essentialized Third World Woman.[14] Subjugation is not grounds for an ontology; it might be a visual clue. Vision requires instruments of vision; an optics is a politics of positioning. Instruments of vision mediate standpoints; there is no immediate vision from the standpoints of the subjugated. Identity, including self-identity, does not produce science; critical positioning does, that is, objectivity. Only those occupying the positions of the dominators are self-identical, unmarked, disembodied, unmediated, transcendent, born again. It is unfortunately possible for the subjugated to lust for and even scramble into that subject position—and then disappear from view. Knowledge from the point of view of the unmarked is truly fantastic, distorted, and irrational. The only position from which objectivity could not possibly be practiced and honored is the standpoint of the master, the Man, the One God, whose Eye produces, appropriates, and orders all difference. No one ever accused the God of monotheism of objectivity, only of indifference. The god trick is self-identical, and we have mistaken that for creativity and knowledge, omniscience even.

Positioning is, therefore, the key practice in grounding knowledge organized around the imagery of vision, and much Western scientific and philosophic discourse is organized in this way. Positioning implies responsibility for our enabling practices. It follows that politics and ethics ground struggles for and contests over what may count as rational knowledge. That is, admitted or not, politics and ethics ground struggles over knowledge projects in the exact, natural, social, and human sciences. Otherwise, rationality is simply impossible, an optical illusion projected from nowhere comprehensively. Histories of science may be powerfully told as histories of the technologies. These technologies are ways of life, social orders, practices of visualization. Technologies are skilled practices. How to see? Where to see from? What limits to vision? What to see for? Whom to see with? Who gets to have more than one point of view? Who gets blinded? Who wears blinders? Who interprets the visual field? What other sensory powers do we wish to cultivate besides vision? Moral and political discourse should be the paradigm for rational discourse about the imagery and technologies of vision. Sandra Harding's claim, or observation, that movements of social revolution have most contributed to improvements in science might be read as a claim about the knowledge consequences of new technologies of positioning. But I wish Harding had spent more time remembering that social and scientific revolutions have not always been liberatory, even if they have always been visionary. Perhaps this point could be captured in another phrase: the science question in the military. Struggles over what will count as rational accounts of the world are struggles over *how* to see. The terms of vision: the science question in colonialism, the science question in exterminism,[15] the science question in feminism.

The issue in politically engaged attacks on various empiricisms, reductionisms, or other versions of scientific authority should not be relativism—but location. A dichotomous chart expressing this point might look like this:

universal rationality	ethnophilosophies
common language	heteroglossia
new organon	deconstruction
unified field theory	oppositional positioning
world system	local knowledges
master theory	webbed accounts

But a dichotomous chart misrepresents in a critical way the positions of embodied objectivity that I am trying to sketch. The primary distortion is the illusion of symmetry in the chart's dichotomy, making any position appear, first, simply alternative and, second, mutually exclusive. A map of tensions and resonances between the fixed ends of a charged dichotomy better represents the potent politics and epistemologies of embodied, therefore accountable, objectivity. For example, local knowledges have also to be in tension with the productive structurings that force unequal translations and exchanges—material and semiotic—within the webs of knowledge and power. Webs *can* have the property of being systematic, even of being centrally structured global systems with deep filaments and tenacious tendrils into time, space, and consciousness, which are the dimensions of world history. Feminist accountability requires a knowledge tuned to resonance, not to dichotomy. Gender is a field of structured and structuring difference, in which the tones of extreme localization, of the intimately personal and individualized body, vibrate in the same field with global high-tension emissions. Feminist embodiment, then, is not about fixed location in a reified body, female or otherwise, but about nodes in fields, inflections in orientations, and responsibility for difference in material-semiotic fields of meaning. Embodiment is significant prosthesis; objectivity cannot be about fixed vision when what counts as an object is precisely what world history turns out to be about.

How should one be positioned in order to see, in this situation of tensions, resonances, transformations, resistances, and complicities? Here, primate vision is not immediately a very powerful metaphor or technology for feminist political-epistemological clarification, because it seems to present to consciousness already processed and objectified fields: things seem already fixed and distanced. But the visual metaphor allows one to go beyond fixed appearances, which are only the end products. The metaphor invites us to investigate the varied apparatuses of visual production, including the prosthetic technologies interfaced with our biological eyes and brains. And here we find highly particular machineries for processing regions of the electromagnetic spectrum into our pictures of the world. It is in the intricacies of these visualization technologies in which we are embedded that we will find metaphors and means for understanding and intervening in the patterns of objectification in the world—that is, the patterns of reality for which we must be accountable. In these metaphors, we find means for appreciating simultaneously *both* the concrete, "real" aspect and the aspect of semiosis and production in what we call scientific knowledge.

I am arguing for politics and epistemologies of location, positioning, and situating, where partiality and not universality is the condition of being heard to make rational knowledge claims. These are claims on people's lives. I am arguing for the view from a body, always a complex, contradictory, structuring, and structured body, versus the view from above, from nowhere, from simplicity. Only the god trick is forbidden. Here is a criterion for deciding the science question in militarism, that dream science/technology of perfect language, perfect communication, final order.

Feminism loves another science: the sciences and politics of interpretation, translation, stuttering, and the partly understood. Feminism is about the sciences of the multiple subject with (at least) double vision. Feminism is about a critical vision consequent upon a critical positioning in unhomogeneous gendered social space.[16] Translation is always interpretive, critical, and partial. Here is a ground for conversation, rationality, and objectivity—which is power-sensitive, not pluralist, "conversation." It is not even the mythic cartoons of physics and mathematics—incorrectly caricatured in antiscience ideology as exact, hypersimple knowledges—that have come to represent the hostile other to feminist paradigmatic models of scientific knowledge, but the dreams of the perfectly known in high-technology, permanently militarized scientific productions and positionings, the god trick of a Star Wars paradigm of rational knowledge. So location is about vulnerability; location resists the politics of closure, finality, or to borrow from Althusser, feminist objectivity resists "simplification in the last instance." That is because feminist embodiment resists fixation and is insatiably curious about the webs of differential positioning. There is no single feminist standpoint because our maps require too many dimensions for that metaphor to ground our visions. But the feminist standpoint theorists' goal of an epistemology and politics of engaged, accountable positioning remains eminently potent. The goal is better accounts of the world, that is, "science."

Above all, rational knowledge does not pretend to disengagement: to be from everywhere and so nowhere, to be free from interpretation, from being represented, to be fully self-contained or fully formalizable. Rational knowledge is a process of ongoing critical interpretation among "fields" of interpreters and decoders. Rational knowledge is power-sensitive conversation.[17] Decoding and transcoding plus translation and criticism; all are necessary. So science becomes the paradigmatic model, not of closure, but of that which is contestable and contested. Science becomes the myth, not of what escapes human agency and responsibility in a realm above the fray, but, rather, of accountability and responsibility for translations and solidarities linking the cacophonous visions and visionary voices that characterize the knowledges of the subjugated. A splitting of senses, a confusion of voice and sight, rather than clear and distinct ideas, becomes the metaphor for the ground of the rational. We seek not the knowledges ruled by phallogocentrism (nostalgia for the presence of the one true Word) and disembodied vision. We seek those ruled by partial sight and limited voice—not partiality for its own sake but, rather, for the sake of the connections and unexpected openings situated knowledges make possible. Situated knowledges are about communities, not about isolated individuals. The only way to find a larger vision is to be somewhere in particular. The science question in feminism is about objectivity as positioned rationality. Its images are not the products of escape and transcendence of limits (the view from above) but the joining of partial views and halting voices into a collective subject position that promises a vision of the means of ongoing finite embodiment, of living within limits and contradictions—of views from somewhere.

Objects as Actors: The Apparatus of Bodily Production

Throughout this reflection on "objectivity," I have refused to resolve the ambiguities built into referring to science without differentiating its extraordinary range of contexts. Through the insistent ambiguity, I have foregrounded a field of commonalities binding exact, physical, natural, social, political, biological, and human sciences; and I have tied this whole heterogeneous field of academically (and industrially, e.g., in publishing, the weapons trade, and pharmaceuticals) institutionalized knowledge production to a meaning of science that insists on its potency in ideological struggles. But, partly in order to give play to both the specificities and the highly permeable boundaries of meanings in discourse on science, I would like to suggest a resolution to one ambiguity. Throughout the field of meanings constituting science, one of the commonalities concerns the status of any object of knowledge and of related claims about the faithfulness of our accounts to a "real world," no matter how mediated for us and no matter how complex and contradictory these worlds may be. Feminists, and others who have been most active as critics of the sciences and their claims or associated ideologies, have shied away from doctrines of scientific objectivity in part because of the suspicion that an "object" of knowledge is a passive and inert thing. Accounts of such objects can seem to be either appropriations of a fixed and determined world reduced to resource for instrumentalist projects of destructive Western societies, or they can be seen as masks for interests, usually dominating interests.

For example, "sex" as an object of biological knowledge appears regularly in the guise of biological determinism, threatening the fragile space for social constructionism and critical theory, with their attendant possibilities for active and transformative intervention, which were called into being by feminist concepts of gender as socially, historically, and semiotically positioned difference. And yet, to lose authoritative biological accounts of sex, which set up productive tensions with gender, seems to be to lose too much; it seems to be to lose not just analytic power within a particular Western tradition but also the body itself as anything but a blank page for social inscriptions, including those of biological discourse. The same problem of loss attends the radical "reduction" of the objects of physics or of any other science to the ephemera of discursive production and social construction.[18]

But the difficulty and loss are not necessary. They derive partly from the analytic tradition, deeply indebted to Aristotle and to the transformative history of "White Capitalist Patriarchy" (how may we name this scandalous Thing?) that turns everything into a resource for appropriation, in which an object of knowledge is finally itself only matter for the seminal power, the act, of the knower. Here, the object both guarantees and refreshes the power of the knower, but any status as *agent* in the productions of knowledge must be denied the object. It—the world—must, in short, be objectified as a thing, not as an agent; it must be matter for the self-formation of the only social being in the productions of knowledge, the human knower. Zoe Sofoulis[19] identified the structure of this mode of knowing in technoscience as "resourcing"—as the second-birthing of Man through the homogenizing of all the world's body into resource for his perverse projects. Nature is only the raw material of culture, appropriated, preserved, enslaved, exalted, or otherwise made flexible for disposal by culture in the logic of capitalist colonialism. Similarly, sex is only matter to the act of gender; the productionist logic seems inescapable in traditions of Western binary oppositions. This analytical and historical narrative logic accounts for my nervousness about the sex/gender distinction in the recent history of feminist theory. Sex is "resourced" for its representation as gender, which "we" can control. It has seemed all but impossible to avoid the trap of an appropriationist logic of domination built into the nature/culture opposition and its generative lineage, including the sex/gender distinction.

It seems clear that feminist accounts of objectivity and embodiment—that is, of a world—of the kind sketched in this essay require a deceptively simple maneuver within inherited Western analytical traditions, a maneuver begun in dialectics but stopping short of the needed revisions. Situated knowledges require that the object of knowledge be pictured as an actor and agent, not as a screen or a ground or a resource, never finally as slave to the master that closes off the dialectic in his unique agency and his authorship of "objective" knowledge. The point is paradigmatically clear in critical approaches to the social and human sciences, where the agency of people studied itself transforms the entire project of producing social theory. Indeed, coming to terms with the agency of the "objects" studied is the only way to avoid gross error and false knowledge of many kinds in these sciences. But the same point must apply to the other knowledge projects called sciences. A corollary of the insistence that ethics and politics covertly or overtly provide the bases for objectivity in the sciences as a heterogeneous whole, and not just in the social sciences, is granting the status of agent/actor to the "objects" of the world. Actors come in many and wonderful forms. Accounts of a "real" world do not, then, depend on a logic of "discovery" but on a power-charged social relation of "conversation." The world neither speaks itself nor disappears in favor of a master decoder. The codes of the world are not still, waiting only to be read. The world is not raw material for humanization; the thorough attacks on humanism, another branch of "death of the subject" discourse, have made this point quite clear. In some critical sense that is crudely hinted at by the clumsy category of the social or of agency, the world encountered in knowledge projects is an active entity. Insofar as a scientific account has been able to engage this dimension of the world as object of knowledge, faithful knowledge can be imagined and can make claims on us. But no particular doctrine of representation or decoding or discovery guarantees anything. The approach I am recommending is not a version of "realism," which has proved a rather poor way of engaging with the world's active agency.

My simple, perhaps simpleminded, maneuver is obviously not new in Western philosophy, but it has a special feminist edge to it in relation to the science question in feminism and to the linked question of gender as situated difference and the question of female embodiment. Ecofeminists have perhaps been most insistent on some version of the world as active subject, not as resource to be mapped and appropriated in bourgeois, Marxist, or masculinist projects. Acknowledging the agency of the world in knowledge makes room for some unsettling possibilities, including a sense of the world's independent sense of humor. Such a sense of humor is not comfortable for humanists and others committed to the world as resource. There are, however, richly evocative figures to promote feminist visualizations of the world as witty agent. We need not lapse into appeals to a primal mother resisting her translation into resource. The Coyote or Trickster, as embodied in Southwest native American accounts, suggests the situation we are in when we give up mastery but keep searching for fidelity, knowing all the while that we will be hoodwinked. I think these are useful myths for scientists who might be our allies. Feminist objectivity makes room for surprises and ironies at the heart of all knowledge production; we are not in charge of the world. We just live here and try to strike up noninnocent conversations by means of our prosthetic devices, including our visualization technologies. No wonder science fiction has been such a rich writing practice in recent feminist theory. I like to see feminist theory as a reinvented coyote discourse obligated to its sources in many heterogeneous accounts of the world.

Another rich feminist practice in science in the last couple of decades illustrates particularly well the "activation" of the previously passive categories of objects of knowledge. This activation permanently problematizes binary distinctions such as sex and gender, without eliminating their strategic utility. I refer to the reconstructions in primatology (especially, but not only, in women's practice as primatologists, evolutionary biologists, and behavioral ecologists) of what may count as sex, especially as female sex, in scientific accounts.[20] The *body*, the object of biological discourse, becomes a most engaging being. Claims of biological determinism can never be the same again. When female "sex" has been so thoroughly retheorized and revisualized that it emerges as practically indistinguishable from "mind," something basic has happened to the categories of biology. The biological female peopling current biological behavioral accounts has almost no passive properties left. She is structuring and active in every respect; the "body" is an agent, not a resource. Difference is theorized *biologically* as situational, not intrinsic, at every level from gene to foraging pattern, thereby fundamentally changing the biological politics of the body. The relations between sex and gender need to be categorically reworked within these frames of knowledge. I would like to suggest that this trend in explanatory strategies in biology is an allegory for interventions faithful to projects of feminist objectivity. The point is not that these new pictures of the biological female are simply true or not open to contestation and conversation—quite the opposite. But these pictures foreground knowledge as situated conversation at every level of its articulation. The boundary between animal and human is one of the stakes in this allegory, as is the boundary between machine and organism.

So I will close with a final category useful to a feminist theory of situated knowledges: the apparatus of bodily production. In her analysis of the production of the poem as an object of literary value, Katie King offers tools that clarify matters in the objectivity debates among feminists. King suggests the term *apparatus of literary production* to refer to the emergence of literature at the intersection of art, business, and technology. The apparatus of literary production is a matrix from which "literature" is born. Focusing on the potent object of value called the "poem," King applies her analytic framework to the relation of women and writing technologies.[21] I would like to adapt her work to understanding the generation—the actual production and reproduction—of bodies and other objects of value in scientific knowledge projects. At first glance, there is a limitation to using King's scheme inherent in the "facticity" of biological discourse that is absent from literary discourse and its knowledge claims. Are biological bodies "produced" or "generated" in the same strong sense as poems? From the early stirrings of romanticism in the late eighteenth century, many poets and biologists have believed that poetry and organisms are siblings. *Frankenstein* may be read as a meditation on this proposition. I continue to believe in this potent proposition but in a postmodern and not a romantic manner. I wish to translate the ideological dimensions of "facticity" and "the organic" into a cumbersome entity called a "material-semiotic actor." This unwieldy term is intended to portray the object of knowledge as an active, meaning-generating part of apparatus of bodily production, without *ever* implying the immediate presence of such objects or, what is the same thing, their final or unique determination of what can count as objective knowledge at a particular historical juncture. Like "poems," which are sites of literary production where language too is an actor independent of intentions and authors, bodies as objects of knowledge are material-semiotic generative nodes. Their *boundaries* materialize in social interaction. Boundaries are drawn by mapping practices; "objects" do not preexist as such. Objects are boundary projects. But boundaries shift from within; boundaries are very tricky. What boundaries provisionally contain remains generative, productive of meanings and bodies. Siting (sighting) boundaries is a risky practice.

Objectivity is not about disengagement but about mutual *and* usually unequal structuring, about taking risks in a world where "we" are permanently mortal, that is, not in "final" control. We have, finally, no clear and distinct ideas. The various contending biological bodies emerge at the intersection of biological research and writing, medical and other business practices, and technology, such as the visualization tech-nologies enlisted as metaphors in this essay. But also invited into that node of intersec-tion is the analogue to the lively languages that actively intertwine in the production of literary value: the coyote and the protean embodiments of the world as witty agent and actor. Perhaps the world resists being reduced to mere resource because it is—not mother/matter/mutter—but coyote, a figure of the always problematic, always potent tie between meaning and bodies. Feminist embodiment, feminist hopes for partiality, objectivity, and situated knowledges, turn on conversations and codes at this potent node in fields of possible bodies and meanings. Here is where science, science fantasy, and science fiction converge in the objectivity question in feminism. Perhaps our hopes for accountability, for politics, for ecofeminism, turn on revisioning the world as coding trickster with whom we must learn to converse.

Notes and References

This chapter originated as a commentary on Sandra Harding's *The Science Question in Feminism*, at the Western Division meetings of the American Philosophical Association, San Francisco, March 1987. Support during the writing of this paper was generously provided by the Alpha Fund of the Institute for Advanced Study, Princeton, New Jersey. Thanks especially to Joan Scott, Judy Butler, Lila Abu-Lughod, and Dorinne Kondo.

1. For example, see Karin Knorr-Cetina and Michael Mulkay, eds., *Science Observed: Per-spectives on the Social Study of Science* (London: Sage, 1983); Wiebe B. Bijker, Thomas P. Hughes, and Trevor Pinch, eds., *The Social Construction of Technological Systems* (Cambridge, MA: MIT Press, 1987); and esp. Bruno Latour's *Les microbes, guerre et paix, suivi d'irréductions* (Paris: Métailié, 1984) and *The Pasteurization of France, Followed by Irreductions: A Politico-Scientific Essay* (Cambridge, MA: Harvard University Press, 1988). Borrowing from Michel Tournier's *Ven-dredi* (Paris: Gallimard, 1967), *Les microbes* (p. 171), Latour's brilliant and maddening aphoristic polemic against all forms of reductionism, makes the essential point for feminists: "Méfiez-vous de la pureté; c'est le vitriol de l'âme" (Beware of purity; it is the vitriol of the soul). Latour is not other-wise a notable feminist theorist, but he might be made into one by readings as perverse as those he makes of the laboratory, that great machine for making significant mistakes faster than anyone else can, and so gaining world-changing power. The laboratory for Latour is the railroad industry of epistemology, where facts can be made to run on the tracks laid down from the laboratory out. Those who control the railroads control the surrounding territory. How could we have forgotten? But now it's not so much the bankrupt railroads we need as the satellite network. Facts run on light beams these days.

2. For an elegant and very helpful elucidation of a noncartoon version of this argument, see Hayden White, *The Content of the Form: Narrative Discourse and Historical Representation* (Balti-more: Johns Hopkins University Press, 1987). I still want more; and unfulfilled desire can be a pow-erful seed for changing the stories.

3. In "Through the Lumen: Frankenstein and the Optics of Re-Origination" (Ph.D. diss. Uni-versity of California at Santa Cruz, 1988), Zoe Sofoulis has produced a dazzling (she will forgive me the metaphor) theoretical treatment of technoscience, the psychoanalysis of science-fiction culture,

and the metaphorics of extraterrestrialism, including a wonderful focus on the ideologies and philosophies of light, illumination, and discovery in Western mythics of science and technology. My essay was revised in dialogue with Sofoulis's arguments and metaphors in her dissertation.

4. Nancy Hartsock, *Money, Sex, and Power: An Essay on Domination and Community* (Boston: Northeastern University Press, 1984).

5. Crucial to this discussion are Sandra Harding, *The Science Question in Feminism* (Ithaca: Cornell University Press, 1987); Evelyn Fox Keller, *Reflections on Gender and Science* (New Haven: Yale University Press, 1984); Nancy Hartsock, "The Feminist Standpoint: Developing the Ground for a Specifically Feminist Historical Materialism," in *Discovering Reality: Feminist Perspectives on Epistemology, Metaphysics, and Philosophy of Science*, ed. Sandra Harding and Merrill B. Hintikka (Dordrecht, The Netherlands: Reidel, 1983): 283–310; Jane Flax's "Political Philosophy and the Patriarchal Unconscious," in *Discovering Reality*, pp. 245–81; and "Postmodernism and Gender Relations in Feminist Theory," *Signs* 12 (Summer 1987): 621–43; Evelyn Fox Keller and Christine Grontkowski, "The Mind's Eye," in *Discovering Reality*, pp. 207–24; Hilary Rose, "Women's Work, Women's Knowledge," in *What Is Feminism? A Re-Examination*, ed. Juliet Mitchell and Ann Oakley (New York: Pantheon, 1986), pp. 161–83; Donna Haraway, "A Manifesto for Cyborgs: Science, Technology, and Socialist Feminism in the 1980s," *Socialist Review*, no. 80 (March–April 1985): 65–107; and Rosalind Pollack Petchesky, "Fetal Images: The Power of Visual Culture in the Politics of Reproduction," *Feminist Studies* 13 (Summer 1987): 263–92.

Aspects of the debates about modernism and postmodernism affect feminist analyses of the problem of "objectivity." Mapping the fault line between modernism and postmodernism in ethnography and anthropology—in which the high stakes are the authorization or prohibition to craft *comparative* knowledge across "cultures"—Marilyn Strathern made the crucial observation that it is not the written ethnography that is parallel to the work of art as object-of-knowledge, but the *culture*. The romantic and modernist natural-technical objects of knowledge, in science and in other cultural practice, stand on one side of this divide. The postmodernist formation stands on the other side, with its "anti-aesthetic" of permanently split, problematized, always receding and deferred "objects" of knowledge and practice, including signs, organisms, systems, selves, and cultures. "Objectivity" in a postmodern framework cannot be about unproblematic *objects*; it must be about specific prosthesis and always partial translations. At root, objectivity is about crafting *comparative* knowledge: How may a community name things to be stable and to be like each other? In postmodernism, this query translates into a question of the politics of redrawing of boundaries in order to have non-innocent conversations and connections. What is at stake in the debates about modernism and postmodernism is the pattern of relationships between and within bodies and language. This is a crucial matter for feminists. See Marilyn Strathern, "Out of Context: The Persuasive Fictions of Anthropology," *Current Anthropology* 28 (June 1987): 251–81, and "Partial Connections," Munro Lecture, University of Edinburgh, November 1987, unpublished manuscript.

6. Harding, pp. 24–26, 161–62.

7. John Varley's science-fiction short story "The Persistence of Vision," in *The Persistence of Vision* (New York: Dell, 1978), pp. 263–316, is part of the inspiration for this section. In the story, Varley constructs a utopian community designed and built by the deaf-blind. He then explores these people's technologies and other mediations of communication and their relations to sighted children and visitors. In the story "Blue Champagne," in *Blue Champagne* (New York: Berkeley, 1986), pp. 17–79, Varley transmutes the theme to interrogate the politics of intimacy and technology for a paraplegic young woman whose prosthetic device, the golden gypsy, allows her full mobility. But because the infinitely costly device is owned by an intergalactic communications and entertainment empire, for which she works as a media star making "feelies," she may keep her technological, intimate, enabling, other self only in exchange for her complicity in the commodification of all experience. What are her limits to the reinvention of experience for sale? Is the personal political under the sign of simulation? One way to read Varley's repeated investigations of finally always limited embodiments, differently abled beings, prosthetic technologies, and cyborgian encounters with their finitude, despite their extraordinary transcendence of "organic" orders, is to find an allegory for the personal and political in the historical mythic time of the late twentieth century, the era

of techno-biopolitics. Prosthesis becomes a fundamental category for understanding our most inti-mate selves. Prosthesis is semiosis, the making of meanings and bodies, not for transcendence, but for power-charged communication.

8. C. D. B. Bryan, *The National Geographic Society: 100 Years of Adventure and Discovery* (New York: Harry N. Abrams, 1987), p. 352.

9. I owe my understanding of the experience of these photographs to Jim Clifford, University of California at Santa Cruz, who identified their "land ho!" effect on the reader.

10. Bryan, p. 454.

11. See Hartsock, "The Feminist Standpoint: Developing the Ground for a Specifically Femi-nist Historical Materialism"; and Chela Sandoral, *Yours in Struggle: Women Respond to Racism* (Oakland: Center for Third World Organizing, n.d.); Harding; and Gloria Anzaldua, *Borderlands/La Frontera* (San Francisco: Spinsters/Aunt Lute, 1987).

12. Annette Kuhn, *Women's Pictures: Feminism and Cinema* (London: Routledge & Kegan Paul, 1982), pp. 3–18.

13. Joan Scott reminded me that Teresa de Lauretis put it like this:

Differences among women may be better understood as differences within women. . . . But once understood in their constitutive power—once it is understood, that is, that these differences not only constitute each woman's consciousness and subjective limits but all together define the *female subject of feminism* in its very specificity, in inherent and at least for now irreconcilable contradiction—these differences, then, cannot be again collapsed into a fixed identity, a same-ness of all women as Woman, or a representation of Feminism as a coherent and available image.

See Teresa de Lauretis, "Feminist Studies/Critical Studies: Issues, Terms, and Contexts," in her *Feminist Studies/Critical Studies* (Bloomington: Indiana University Press, 1986), pp. 14–15.

14. Chandra Mohanty, "Under Western Eyes," *Boundary* 2 and 3 (1984): 333–58.

15. See Sofoulis, unpublished manuscript.

16. In *The Science Question in Feminism* (p. 18), Harding suggests that gender has three di-mensions, each historically specific: gender symbolism, the social-sexual division of labor, and processes of constructing individual gendered identity. I would enlarge her point to note that there is no reason to expect the three dimensions to covary or codetermine each other, at least not directly. That is, extremely steep gradients between contrasting terms in gender symbolism may very well not correlate with sharp social-sexual divisions of labor or social power, but they may be closely related to sharp racial stratification or something else. Similarly, the processes of gendered subject formation may not be directly illuminated by knowledge of the sexual division of labor or the gender symbolism in the particular historical situation under examination. On the other hand, we should expect mediated relations among the dimensions. The mediations might move through quite differ-ent social axes of organization of both symbols, practice, and identity, such as race—and vice versa. I would suggest also that science, as well as gender or race, might be usefully broken up into such a multipart scheme of symbolism, social practice, and subject position. More than three dimensions suggest themselves when the parallels are drawn. The different dimensions of, for example, gender, race, and science might mediate relations among dimensions on a parallel chart. That is, racial divisions of labor might mediate the patterns of connection between symbolic connections and formation of individual subject positions on the science or gender chart. Or formations of gendered or racial subjectivity might mediate the relations between scientific social division of labor and scientific symbolic patterns.

The chart below begins an analysis by parallel dissections. In the chart (and in reality?), both gender and science are analytically asymmetrical; that is, each term contains and obscures a struc-turing hierarchicalized binary opposition, sex/gender and nature/science. Each binary opposition orders the silent term by a logic of appropriation, as resource to product, nature to culture, potential to actual. Both poles of the opposition are constructed and structure each other dialectically. Within each voiced or explicit term, further asymmetrical splittings can be excavated, as from gender, masculine to feminine, and from science, hard sciences to soft sciences. This is a point about remembering how a particular analytical tool works, willy-nilly, intended or not. The chart reflects common ideological aspects of discourse on science and gender and may help as an analytical tool to crack open mystified units like Science or Woman.

GENDER	SCIENCE
1. symbolic system	symbolic system
2. social division of labor (by sex, by race, etc.)	social division of labor (e.g., by craft or industrial logics)
3. individual identity/subject position (desiring/desired; autonomous relational)	individual identity/subject position (knower/known; scientist/other)
4. material culture (e.g., gender paraphernalia and daily gender technologies, the narrow tracks on which sexual difference runs)	material culture (e.g., laboratories, the narrow tracks on which facts run)
5. dialect of construction and discovery	dialectic of construction and discovery

17. Katie King, "Canons without Innocence" (Ph.D. diss., University of California at Santa Cruz, 1987).

18. Evelyn Fox Keller, in "The Gender/Science System: Or, Is Sex to Gender as Nature Is to Science?" (*Hypatia* 2 [Fall 1987]: 37–49), has insisted on the important possibilities opened up by the construction of the intersection of the distinction between sex and gender, on the one hand, and nature and science, on the other. She also insists on the need to hold to some nondiscursive grounding in "sex" and "nature," perhaps what I am calling the "body" and "world."

19. See Sofoulis, chapter 3.

20. Donna Haraway, *Primate Visions: Gender, Race, and Nature in the World of Modern Science* (New York: Routledge & Kegan Paul), 1989.

21. Katie King, prospectus for "The Passing Dreams of Choice . . . Once Before and After: Audre Lorde and the Apparatus of Literary Production" (*MS*, University of Maryland, College Park, Maryland, 1987).

12

Knowledge, Bodies, and Values

Reproductive Technologies and Their Scientific Context

Helen E. Longino

I T IS POSSIBLE to set and evaluate human reproductive technologies in many contexts: their current contexts of implementation; the context of traditional values as defined, for example, by the Vatican or recent United States administrations; the context of progressive values associated with environmentalism, population control, and women's emancipation, the context of medical technologies, the context of biotechnology in general. Each contextual setting would yield a different pattern of connections. I propose in this chapter to think about these technologies against the background of biological research exploiting the discovery of the structure of the DNA molecule in the early 1950s. Fundamental research in molecular biology has been intimately connected with its technological applications—so much so that the line between pure and applied science has all but vanished. The current science studies expression "technoscience" seems particularly apt for the union of molecular biology, genetic engineering, and procreative technologies that this contextualization highlights. By setting reproductive technologies in this research context and then setting the research in the framework of a philosophical analysis of the role of social values in scientific inquiry I hope to offer a perspective on these technologies that is relevant to the concerns of their social critics.

I

In the mid-1940s, as Evelyn Keller and other historians tell us, many physicists left physics for biology to pursue the secrets not of death, which they had managed to display so spectacularly in the air over Hiroshima and Nagasaki, but of life.[1] They believed these secrets to be hidden in the genetic material of organisms. The intense research effort that followed experienced its first culmination with the publication of James Watson and Francis Crick's letter to *Nature* in 1953, describing the molecular structure of the DNA molecule.[2] As Watson and Crick coyly noted, the double helical structure with its matched pairs of bases provided the answer to the question how genetic information could be transmitted from parent cell to daughter cell and from parent organism to offspring. When the two helices of the molecule unwind, each constitutes a template for the formation of a new strand, the new and the template to be wound together in a new double helical DNA molecule. Watson and Crick had provided the alphabet and, by specifying the sequence of bases and pairings, some of the formulation rules for the language of DNA and by extension for the language of biological heredity. It seemed to remain only to decipher the words and sentences inscribed in the molecule—to figure out what constituted a discrete sequence—a codon—carrying a discrete instruction.

In the following decades, molecular biologists proceeded to learn the grammar and syntax of the bases. This task proceeded hand in hand with its technological applications in so-called genetic engineering, first in simpler bacterial life forms, later in more complicated forms, multicellular organisms composed of eukaryotic cells. The ability to identify sequences of nucleotides with the synthesis of particular proteins and the development of techniques to cut specific nucleotide sequences from the DNA molecule and insert them into another molecule enabled the molecular biologists to get bacterial plasmids to produce a number of useful biological substances. (Some, however, like interferon, are still "miracle drugs in search of a disease.") This research exploited the reproductive capacities of naturally occurring entities to produce material in much greater quantities than otherwise available. It thus spawned the growth of a biotechnology industry and made a number of biologists very wealthy. The productive/reproductive technology and science moved from bacteria, to plants, to animals, and aspires, finally, to include humans. Whether using a bacterially synthesized bioproduct like insulin to help someone control their diabetes, trying to produce higher yielding dairy cows with bacterially produced bovine growth hormone, or hoping to cure a genetic disease by gene transplantation, biologists have been expanding their control over life processes via their ability to read and reproduce the text of the DNA molecule.

Because the technological applications have developed as part of the basic science, and some results have, therefore, come so quickly, the public image of biotechnology is that of having achieved even greater mastery over biological processes than it actually has. But knowledge of the grammar and syntax of a language does not explain instances of its acquisition and use. As will be clearer below, we are quite far from knowing what traits are correlated with what (sets of) gene sequences, and far from understanding how genes are activated.

Enough has been accomplished, however, to stimulate vigorous debates over the ethical and social dimensions of uncontrolled development of biotechnology. Concerns are raised about the effects of genetically engineered crops on other species, their effects on those who consume them. At the UN conference on the Environment in Rio de Janeiro, the issue of who can claim ownership of genetic material prevented the United States from signing a treaty on biodiversity. And if these technologies can be extended to humans, who will decide how they are to be used?

These questions take specific form in the debate over reproductive technologies. While the latter include a wide range of developments from inhibitory technologies such as the development of birth control methods utilizing synthetic hormones to productive technologies, I shall concentrate on the latter. These range from relatively low-tech methods such as artificial insemination to the high-tech and high-profile in vitro fertilization, with its associated technologies of egg-harvesting and embryo or zygote implantation as well as embryo transfer. Many of these technologies were first developed in the context of animal breeding. If one extends the category to include medical technologies related to reproduction, one would also include amniocentesis, chorionic villus sampling, ultrasound, other forms of fetal monitoring, and the preservation of prematurely delivered infants. The ability to keep embryos resulting from in vitro fertilization alive for a certain period of time together with the ability to extend the viability of premature infants ever earlier into the gestational period makes the prospect of an artificial womb gestating a fetus resulting from preserved engineered germ cells every day less science-fictional and more realistic.

These technologies, like biotechnologies in general, are double-edged, and feminists have been divided in their assessments of them. While few feminists defend them unequivocally, some defend them as potentially offering women greater control over their reproductive lives. Fetal monitoring techniques seem to offer women greater certainty about the health of the fetus and its prospects once born.[3] In vitro fertilization and embryo implantation enable infertile women to bear children if they wish to.

In spite of these benefits, even defenders of the technologies acknowledge problems with present forms of implementation.[4] Just as contraceptive technologies were introduced in ways that benefited some and disadvantaged others, so the reproductive technologies are not equally available to all. In vitro fertilization is still a very expensive procedure, and hence limited to those with the means to pay for it. But other restrictions are also placed on access to the technologies. These vary from country to country, and also from clinic to clinic. In Sweden, for example, even artificial insemination is limited to women in traditional heterosexual marriages, and denied to lesbians. This is in contrast to the situation in the United States, where artificial insemination has been widely practiced (in clinical and non-clinical settings) by lesbians and single heterosexual women wishing to bear children. Access to in vitro fertilization is more widely restricted to those who fit what have been thought to be appropriate criteria for motherhood: heterosexuality, marriage, absence of disabilities. These patterns of restriction and access reinforce traditional conceptions of women's place in society.

In addition to these distributional concerns feminists have been concerned about the well-being of those women who do utilize these technologies, and whether sufficient information is communicated to potential clients in their promotion and adequate attention paid to what might be called the side effects of utilizing them. The procedures of in vitro fertilization are themselves physically and emotionally stressful. The drugs—synthetic hormones—used to stimulate ovulation produce a variety of other effects, including headaches, mood swings, nausea, and dizziness. The daily physical routine is very demanding (ultrasound requires a full bladder), producing fatigue and stress. In addition, the practice of multiple embryo implantation frequently results in multiple births, producing further stress for a couple who might have been prepared for one child, but not for three or four! Finally, the success rate is very low—some estimate it at 5 to 10 percent, although new techniques are raising the rate somewhat. In one study of IVF parents in Canada in the late 1980s, twenty women underwent a total of fifty attempts with an outcome of six infants; three of whom were triplets. Four women out of twenty were successful, four attempts out of fifty were successful.

The promotion of these technologies has occasioned other criticism. Underlying their celebration as an answer to infertility is the assumption that a woman's fulfillment consists in motherhood. IVF simply permits a woman to attain a fulfillment denied her by an accident of nature. Biologist Ruth Doell has argued that while this pronatalism revives stereotypes briefly discredited by the women's movement, we should not read it as expressing biotechnologists' own intentions with respect to the technologies.[5] Rather, she says, biologists are exploiting the pronatalism of the culture for their own ends. The technologies of in vitro fertilization not only assist women who could not otherwise conceive a child to do so, they also make available research materials for the biologist: human eggs and embryos. Her observations remind us that the development of these technologies is a function of a confluence of interests: the research interests of biologists, the pronatalism still effective in the culture, and the individual interests both of the specialists operating the clinics and of the infertile women resorting to them.

These criticisms are directed at current forms of implementation and promotion of these technologies. Another more sweeping critical approach addresses the technologies themselves as a threat to women. Aldous Huxley's *Brave New World* depicted a society in which all human reproduction had become the industrial production of humans by means of genetic engineering and conceptive technologies.[6] Many feminists see the joint development of genetic engineering and conceptive technology as a preliminary to gynocide, the elimination of women.[7] This fear stems from at least two deeply interrelated sources: the repeated expression by men of a desire to appropriate the procreative functions of women on the one hand, and the analysis of some feminists that locates women's source of strength in our procreative capacity or maternal identity on the other.

The first of these can be heard echoing from the ambitions of seventeenth-century alchemists to those of the authors of the National Academy of Sciences report *Biology and the Future of Man.* Historians Sally Allen and Joanna Hubbs have studied the writings of the alchemists, finding in them an aspiration to a collaboration between man and nature to be "achieved by denying the independent status of the feminine and by containing and arrogating her creative powers."[8] They quote Michael Maier: "It makes no difference whether the nest is set in place by the hen or the farmer, for the generation of the egg is the same."[9] But, he goes on to explain, creation by artificial means can eliminate the impurities of the naturally generated and is thus superior to natural creation. How different is this from the stirring peroration to the Academy report? "Now man can guide his own evolution. In him, nature has reached beyond the hard regularities of physical phenomena. *Homo sapiens,* the creature of Nature has transcended her. From a product of circumstances he has risen to responsibility. At last, he is Man. May he behave so!"[10] Man, of course, is more immediately a creature of his mother, and it is the incarnation of Nature in the person of his mother that Man will transcend by means of biological technology.

These masculinist visions are in curious interaction with the second source of the fear of gynocide. Although much of feminist analysis has been devoted to detaching womanhood from motherhood, there is a lingering belief among feminists that women's power and value lie in our procreative capacities.[11] While sometimes represented as our value in and to a male-dominant and sexist society, some feminists also see women's procreative capacity as the source of women's distinctiveness and the basis of female claims on society. If human procreation can be achieved by other means, from this point of view women become redundant.

In light of these and similar concerns, many feminists who share them have argued for the immediate closure of all in vitro fertilization clinics and cessation of research related to conceptive technologies. This, for example, has been the position of the members of FINRRAGE, an international network of feminists organized in response, and for the most part in opposition, to reproductive technologies.[12] Whether or not one agrees with their analysis, it is hard to see the prohibition of research as an objective to furthering any goal. The point is not that freedom of inquiry must be preserved, but that prohibition doesn't work. Too many people want to benefit from reproductive technologies, and too many scientists are eager for the associated research dividends for such a strategy to succeed. Second, this approach grants too much to the science and masks some of the deeper or equally deep issues we might want to raise about it.

II

One of the difficulties in assessing scientific knowledge from a social point of view is our view of science as a mode of value-neutral inquiry. In order to assess scientific research programs at a more comprehensive level than that of their point of contact with human lives in their applications (a point by which they are already well-formed), we need a view of scientific inquiry that can help to reveal the role of social values in the very construction of knowledge. I have tried to develop such a view in my book *Science as Social Knowledge.*[13] I will summarize some of the relevant elements of the view here in order to suggest some directions in which a critical analysis might go.

I call the view I develop contextual empiricism. This is an approach to the analysis of scientific knowledge that joins some traditional features of philosophical empiricism with recent work in philosophy of science and insights from contemporary analysis. It takes from philosophical empiricism the position that sense experience is the fundamental legitimator of (factual) knowledge claims. This is an empiricism that focuses not on the acquisition of belief (or knowledge) but on the justification of factual claims, and that assigns to sense experience, rather than to reason or intention, the role of arbiter among the many claimants for our doxastic allegiance. It also imposes a certain requirement on the content of arguments for factual claims, namely that they specify the observational data available to support such claims. Empiricism, on this interpretation, is a critical doctrine rather than a recipe for or description of knowledge acquisition. The recipe it does offer is one for constructing the logical skeleton of any given scientific argument around which we can then array other elements of the knowledge productive process.

Two features of scientific language and reasoning noted in recent philosophy of science immediately provide some of these other elements. One is the theory-laden character of observation, which when first discussed seemed to threaten all evidential argumentation in science with vicious circularity. As Mary Hesse has argued, however, the theory with which observations are laden may not be the theory the observations are being used to justify.[14] While complicating our view of justificatory arguments in the sciences, theory ladenness, thus, need not be understood as disabling. Nevertheless, in any particular case, data and observation are constituted of sense experience informed by theoretical considerations rather than by "raw" data. Data must be organized and analyzed in order to serve as evidence. A second and related feature of scientific reasoning, first discussed by Pierre Duhem, but taken up by a number of philosophers more recently, is the underdetermination of theory by data.[15] Although sometimes conflated with theory ladenness, this is quite a different logical problem and stems from the semantic independence of theoretical and observational language. The underdetermination results from the discontinuity of language in theoretical claims and that of the observational claims used to ground them, a discontinuity that results from the gap between the level at which our theories purport to describe reality and the level which is accessible to human would-be knowers. This gap is not closed by the theory-laden description of observation. At its most banal, it is the gap between the generality of theoretical statements and the ultimate singularity of observational statements, but it extends to the gap between causal claims and the correlational information that grounds them, and to the gap between theoretical statements about the domains of the very large and the very small relative to human inquirers and what can be said about the mid-sized domain from which their evidence is drawn. It is the gap between statements about galaxies one hundred million light years distant and statements about the measurements made by radio telescopes, between statements about muons and pions and statements about bubble chamber photographs. In these cases the description of the data may well be theory laden, but it does not presuppose the truth of the theoretical claims the data are used to support.

Each of these limitations has a different consequence for our understanding of scientific justificatory structure. As a consequence of theory ladenness, a more prudent empiricism treats sense experience not as the sole legitimator of knowledge claims, since other sorts of considerations may be relevant to establishing the content of observation and the proper articulation of data. These other sorts of consideration, however, are not plucked from the head of Zeus, but are themselves subject to evidential appraisal. Clearly, the justificatory role of experience must be more complexly understood, but the original spirit of empiricism can be retained in the treatment of sense experience as the least defeasible basis of knowledge claims. By this I mean that while every type of claim (theoretical, methodological, observational) used in the justification of a hypothesis can be challenged and is hence corrigible, publicly ascertainable (and ascertained) experiential claims, other things being equal, are accorded greater cognitive authority in the case of conflict.

As a consequence of underdetermination, evidential relations must be understood as constituted by background assumptions which assert relations between the sorts of processes purportedly referred to by theoretical claims and the sorts of phenomena that serve as evidence for them. It follows that a suitable change in background assumption produces a change in evidential status. The same phenomena can support different and even conflicting theoretical hypotheses depending on the context of background assumptions in which their evidential status is assessed. Thus observational data function as the least defeasible bases for hypothesis acceptance within a context that assigns them evidential relevance. In these acknowledgments of both theory ladenness and contextually determined evidential relations, empiricism is modified as contextual empiricism. Experiential data retain their privileged status as the basis of justification, but the description and purported relevance of particular experiences may be corrigible in the light of theoretical considerations, as well as of additional empirical considerations.

While introducing complexity rather than circularity into justificatory reasoning, this qualification of empiricism still threatens traditional ideas about the stability and objectivity of scientific knowledge. One way of expressing the problem is in terms of the distinction between constitutive and contextual values. Constitutive values are those derived from the goals of scientific inquiry. Such goals may be representation, prediction, or understanding of natural phenomena, and they dictate preferences for certain traits in hypotheses. Such traits as empirical adequacy, simplicity, consistency with accepted theory, breadth of scope, etc., may be understood as constitutive values to the extent that they are understood as important to the realization of the goals of inquiry. Contextual values are values pertaining to the social and cultural contexts in which inquiry is carried out, and bear a similar relation to the individual and collective goals of participants in those contexts (some of whom, of course, are participants in scientific inquiry). Contextual values relative to any form of inquiry may include preferences as to who exercises power in a society and how, ideas about the appropriate relation of a given society and culture to others, ideas about interpersonal relationships, about the appropriate objects of worship, etc. How, on the contextual empiricist view, can inquiry be insulated from the influence of contextual values? Background assumptions are carriers of contextual values. If evidential relevance is fixed by mutable background assumptions, what protects knowledge from being the arbitrary expression of subjective desires or the tool of social and personal interests? How can assumptions grounded in contextual values be prevented from serving as background assumptions?

This question has no answer if we remain within an individualist framework, that is, if we think of the producer of knowledge as a single individual. But a number of intellectual movements of the twentieth century, among them feminism, have exposed the illusoriness of individualism and the radical interdependence of human beings in the accomplishment of most tasks. The construction of knowledge is no exception. Understanding the processes of hypothesis justification and acceptance as, fundamentally, processes of social interaction makes possible a solution to the problems of objectivity that remain intractable within an individualist framework. Briefly, the solution is as follows: hypothesis justification involves not only the comparison of hypotheses with observational data, but critical scrutiny of the data, hypotheses, and background assumptions asserting their mutual relevance. Such scrutiny examines claims to accuracy of data reports, the conditions of performance of experiments, the analysis and organization of the data, the contribution of theory to data description, analysis, and organization, the conceptual coherence of the hypotheses, their consistency with current theory, and so forth. By examining the relation between data and theory it identifies the background assumptions constituting their evidential relation and makes them similarly available to scrutiny. Because theoretical and assumptive commitments can be identified as such only from a point of view other than that incorporated in those commitments, this scrutiny is by its nature an intersubjective process, a process that makes sense only in the context of multiple subjectivities. The legitimation and acceptance of scientific hypotheses and theories is, then, a social process, and not an individual one.

Such a critical process assures that idiosyncratic beliefs, values, and interests are excluded from the body of what gets to count as scientific knowledge. Simultaneously, it enables the shaping of this body by socially unquestioned values and interests as expressed in those background assumptions which, in being held by all members of the community, escape criticism. Objectivity, then, as the maximal minimization of subjective (whether individual or collective) preference, is secured through assuring the inclusion of all socially relevant perspectives in the community engaged in the critical construction of knowledge. Only through such inclusion can the community presume that candidates for the status of knowledge have endured adequate critical scrutiny. If knowledge is produced by the community, and if experiential data are not sufficient to validate one hypothesis exclusively against alternatives, then additional constraints must govern the social interactions productive of community acceptance. Such constraints include the availability of forums for critical discourse, uptake of criticism, standards by reference to which criticism is made relevant, and the securing of consensus by dialogue rather than by social or political power.

Objectivity, on this analysis, constitutes an ideal to which communities may aspire, but which they may not necessarily attain. In particular, those communities which, like the United States scientific community, have historically excluded women and members of selected racial and ethnic minorities fail to qualify as objective. This exclusion has meant a marginalization of perspectives that regard women of the dominant race and women and men of excluded racial and ethnic groups as the moral and intellectual equals of those permitted to practice science, and has thereby disabled criticism of theories and hypotheses implicitly or explicitly asserting the opposite.

This analysis contains elements of cognitivist and of constructivist tendencies in contemporary science studies. Scientific knowledge is not simply a matter of apprehending a nature waiting to be discovered. Nature is described and understood through the mediation of assumptions, themselves heterogeneously formed by generalization, analogy, and social and personal aspiration. But neither are our beliefs about the natural world simply a projection of our contextual values. The ultimate test of adequacy is experiential. The natural world imposes constraints on what can (reasonably) be believed of it. Given the gap between our explanatory ambitions and the resources available to satisfy them, the natural world cannot on its own constrain belief to one among all possible alternatives. And while criticism of assumptions, reasoning, and data can further narrow the field, those values and perspectives shared by all members of the community will not be perceived as such, will remain unavailable to criticism, and will shape the knowledge it constructs.

III

How might we apply this analysis to the biological technoscience discussed earlier? There are at least two levels at which contextual values interact with research: development and implementation. With respect to some of the feminist concerns about the impact of the procreative technologies on women, the crucial strategy ought to be to get women fully represented at every level of research and decision making and to ensure a continuing flow of woman-centered analysis to be made available to them. Currently, women are the objects the technology, designed primarily by men, acts upon. If we simply demand a cessation, we remain objects rather than agents. A scientific community deliberately inclusive of those affected by its work might reshape procreative technologies, change the emphases of research, or abandon them altogether. It is not sufficient simply to get a few women into the delivery of technology, as women have generally had to take on a protective coloration when entering male-dominated fields. So long as it is masculine values that women have to absorb in order to be certified in a field, sex is no guarantee of gender identification. It is important, therefore, to ensure not simply the presence of more women, but of women and men who can bring the range of feminist concerns to bear on the development and implementation of procreative technologies.

What seems crucial, at this point, is to sever the research and development of the technologies from their pronatalist public relations. The latter simply obscure the issues for feminists and non-feminists alike. Pronatalism rests on and reinforces an oppressive conception of women's interests and social roles. But it also focuses on the culturally constructed needs of infertile women, particularly if not exclusively women who can afford the costs of assisted conception, drawing social attention away from the reproductive needs of others, e.g., fertile but poor women who cannot obtain adequate prenatal care. It thus perpetuates contemporary Western medicine's focus on expensive, high-technology, interventive cures, rather than low-cost, low-technology prevention. And its parochialism is evident when set in the context of global demographics and the varieties of sexual politics across national and cultural contexts. There is something terribly unbalanced about pouring millions of dollars into an effort to enable a small number of economically privileged women (or the men whose wives they are) to procreate, when there are thousands if not millions of children in need of adoptive or foster homes. If more women of varied cultural, political, and economic backgrounds get more involved in the direction of research, pronatalism should fall away as a red herring, and the scientific and social costs and benefits of each technology may be more realistically assessed.

Second, the analysis enables us to inquire as to the mix of data, theory, and constitutive and contextual values interacting to shape the object of inquiry. For this sort of analysis, it is important to step back from the present configuration of the delivery of these technologies and see them in the larger scientific context within and from which they are emerging. In vitro fertilization, for example, is a necessary condition for the practice of germ cell or embryonic gene replacement therapy. It also provides crucial research materials, eggs and embryos, for research. Gene replacement therapy is itself dependent upon the location of genes responsible for those diseases that are genetically based. Gene therapy engages other values than those engaged by reproductive technologies, and the medical focus switches from cure (for infertility) to prevention (of gene-based diseases), albeit prevention of a decidedly "high tech" sort.

Location of the genes responsible for various diseases is, of course, one of the selling points of the Human Genome Project, whose aim is the mapping of the entire human genome. In a certain number of cases, the specific genetic defect underlying a particular hereditary disease, e.g., cystic fibrosis, has been identified without benefit of the Project's gene mapping. This may explain why other more tantalizing possibilities are dangled before the public. An editorial in *Science* provides a good example of the various assumptions and values involved.[16] Daniel Koshland, the magazine's editor and the editorial's author, simultaneously derided the money spent researching single diseases (specifically cystic fibrosis) and claimed that the Human Genome Project would help unravel multigenic diseases such as heart disease, as well as manic depression, Alzheimer's, and schizophrenia, which he described as "at the root of many social problems." Of the latter, homelessness is the only example Koshland cites. The Human Genome Project would also facilitate identification of those predisposed to "alcoholism, colon cancer, and depression." In the conclusion of the editorial, he urges proceeding with the development of this "great new technology to aid the poor, the infirm, and the underprivileged."[17]

There are a number of points to make concerning these pronouncements. Evelyn Keller has drawn attention to the rhetoric of normal and abnormal which supports the new construal of genetic disease.[18] Since each individual is genetically unique, what will be mapped is a "normal" or standard human genome (or range of genotypes), departures from which constitute abnormalities. But how is "the normal" to be defined and articulated, and by whom? Not by reference to human genes, since determining statistical normality would be far too costly and time-consuming, and would inevitably miss the point, which is to be able to produce effects at the phenotypic level. Thus, "normal" and "abnormal" are categories defined at the phenotypic level and extended by reductive identification to the genotypic level. The extension of the concept of genetic disease to include schizophrenia, alcoholism, and other behavioral phenomena raises the possibility that many other variations in the human condition that give annoyance to others, say chronic lateness, might be classified as "abnormal." Jesting aside,[19] Koshland's reference in all seriousness to such phenomena as homelessness and poverty suggests that, even at the phenotypic level, "normal" is not to be defined statistically, since very large proportions of the world's population qualify as poor (especially by North American standards). "Normal" does not mean normal, but desirable. Its invocation by apologists for the genome project bespeaks not only a desire for uniformity, and perhaps discomfort with the human variety which we daily encounter. It also expresses the alchemist's certainty that he could do a better job than nature of turning out human beings.

Koshland's remarks reveal another dimension of the Human Genome Project, one which it shares with other research projects in contemporary biology. Organisms are conceived in this biological technoscience as products of their genes. While environmental interactions are always acknowledged, they remain peripheral to the conception of development that has until recently been dominant. Humans, too, are conceived as products of our genes. The object of inquiry is the genome as producer of the phenotype. Such an approach individualizes phenomena some of which are and others of which may be social in nature. Poverty and homelessness are certainly social phenomena, both definitionally and etiologically. Koshland's editorial suggests a causal chain from genes to behavioral traits to social conditions, as though such conditions existed apart from a particular context. But both poverty and homelessness are, first of all, defined relative to culturally determined standards. And while some individuals may become poor because of their alcoholism or mental illness, some may become afflicted by these conditions (or find them exacerbated) because of poverty or the loss of a home.[20] Furthermore many, indeed most, poor people are neither alcoholic nor mentally ill; they are simply poor, buffeted by economic forces over which they have no control. Nor is it clear that all instances of mental illness or alcoholism are genetically caused. No research has shown this to anyone's satisfaction; each time a gene is implicated in the manifestation of some mental illness, further research shows the story to be so much more complex as to invalidate any claim of a simple causal relation. A given gene codes either for a contributing factor or for some subvariety of cases, or it turns out to be unrelated.[21]

Not only are these macro-identifications problematic, but recent and contemporary research in development biology suggests that genes are causally effective only in the context of very complex and delicately timed interactions within the cell and organism. So while the presence of a gene may be a sign of a possible given phenotypic condition, its ultimate phenotypic expression depends on many factors apart from the gene. The question is shifting from gene action—how do genes act on cellular material?—to gene activation—how does cellular material selectively engage gene sequences?[22] If the cellular processes in which genes are involved are as complex as current research suggests, then we have given the wrong kind of credit to biotechnologists. The genetic engineers who are able to effect the production of biological substances by inserting the appropriate DNA sequence into a bacterial plasmid or alter the phenotypic traits of plants (e.g., frost resistance or nitrogen fixation) or animals (greater size, higher milk production, etc.) are dependent on and exploit the cell's and organism's ability to function otherwise as normal. Gene manipulation constitutes an opportunistic intervention in complex processes which do not need to be understood in order to function as needed. Just as it was possible to pump water before the role of atmospheric weight and pressure was understood, and to catapult boulders before Newton's laws of motion clarified ballistics, so it has been possible to engineer organisms before biological development is understood. If the worst nightmares of opponents of conceptive technologies presuppose a biological determinism that future research will inevitably undermine, should we not breathe a sigh of relief and turn our attention to matters that do require our intervention?

The relation between scientific inquiry and social values is much more complex than assumed by such a response. Values expressed in inquiry are not simply an outcome of research, nor is their viability in the surrounding culture a function merely of their validation in scientific research programs. Research on the biological basis of behavioral sex differences provides a good example of the interplay between research and social values.[23] This work has been shaped by several interacting sets of values or value-laden assumptions: an ideology of gender dimorphism, biological determinism, and individualism. A variety of theories of male and female behavioral, temperamental, and intellectual differences (almost all positing male superiority) have graced biological thinking in the West since Aristotle. These theories have been discredited either when the general biological view in which they were embedded was replaced by another, or when specific disconfirming research showed them to be erroneous. The remarkable fact about these views is that they have continued to be voiced in whatever categories of biological thought are current, and their disconfirmations have been dismissed as political (and "controversial"). Only recently have researchers in behavior and neurobiology begun to move away from a biologically determinist view of sex-differentiated behaviors to models that posit strong interactions between biology and environment in the development of behavior and the neural circuitry underlying behaviors. This change is contemporaneous with the growing presence and visibility of women in the salaried workforce and in political and cultural life, a presence which is itself a function of economic forces and a strong women's movement. This social change is by no means uniform or uncontested. In the United States, for example, some social policy, for example, welfare policy, is still designed on the assumption that women belong in the home. Job protection for those who take unpaid leave from work for the discharge of parental or filial duties is a matter of debate and was vetoed by the U.S. president. Child care needs are not adequately provided for. The United States Supreme Court is divided on the right of women to reproductive self-determination, and is willing to allow significant restrictions on women's access to abortions. There is no coherent gender ideology, but multiple ideologies in contestation.

The continuing hold of dimorphism and determinism on the cultural and scientific imagination is evinced by the attention given to correlations between behavior and some anatomical or physiological feature and the persistent interpretation of such correlation as evidence of the effect of biological difference on behavioral difference, rather than the reverse.[24] Biological determinism and gender dimorphism are linked to deeply held social values regarding the roles of women and men and the nature of the family. Notions of complex development and of gender variety or singularity are equally linked to social values regarding the emancipation of women, and freedom from gender roles. Thus, the conflicts about the proper structure of society are expressed in the scientific world as disagreement over the nature of gender difference and the degree to which such difference can be detached from biological determinism. The more general idea that people belong in the social categories in which they find themselves, that the wealthy and powerful are so because of inherent (biologically based) abilities or dispositions, is not, however, going the way of gender ideology. The idea remains available to support this ideology should sociopolitical changes revive it, and indeed to support and provide interpretations for those rearguard research programs still engaged in biologizing gender.

This same general idea supports biological determinist interpretations of the relation of genotype and phenotype and their extension to categories of social interest. It is independent of any particular research program and remains alive in the cultural imagination so long as there are significant numbers of people whose interests are served by such an assumption. It draws support from any single determinist research program in progress. However, its independent persistence and opportunistic attachment, both to particular social ideologies (of gender, race, nationality, or social class) on the one hand and to particular research programs (behavioral neuroendocrinology, the biology of group differences in I.Q., etc.) on the other, enable it to survive the collapse of any single element of either and thus to provide an interpretative context for any similar new ideology or research program. This resilience enables it to provide a semblance of "factual" support for social policies, even though the "facts" themselves are constantly being replaced.

What lessons can we learn from these reflections for the future of procreative technologies and the biological research that currently accompanies them? If the history of the interaction of gender ideologies, biological determinism, and biological research is any guide, the slender and decreasing support for simple genetic causality of any phenotypic trait, let alone the very complex conditions mentioned by Koshland, is unlikely to result in any general loss of confidence in the more general thesis of biological determination of socially significant human conditions. Rather this thesis is likely both to find support in temporarily established correlations and to magnify their significance. The danger, therefore, is not that biological research will find a way to make women (or women's reproductive systems) redundant and expendable, nor even that a genuine eugenic program is finally at hand. It is, instead, that the modest successes in the attempts to expand the domain of conceptive research, whether in the technology of conception or the correlation of genes with phenotypic traits, will continue to animate determinist assumptions. There are two consequences. One is that given the interdependence of theoretical and technological developments in molecular biology, reproductive technologies will continue to develop in a way that extends human, especially women's, alienation from the bodily experiences of reproduction. In an ongoing context of sexual inequality, the mastery produced is male mastery. Women and women's bodies are not likely to be expendable, but to continue to be viewed as vessels, props in the ongoing drama of Man's expanding control of Nature. Second, the purported "factual support" means that determinist assumptions can continue as legitimation for social policies that treat the individual as the causal locus of social phenomena.

IV

I have argued in this chapter that reproductive or conceptive technologies are best understood within the more general research context from which they are emerging. Such contextualizing enables us to discern a fuller range of the social values involved in the shaping and promotion of these technologies. It also provides a more moderate view of the possible expansion of human power over nature represented by the current work. Because technologies such as in vitro fertilization are just the visible tip of a more extensive research effort, attempts to curtail it are likely to be unsuccessful. And as a way of effecting social policy and attitudes toward women, repression of the technology would be ineffective. More effective is a coordinated strategy of supporting alternative values in the culture (values of egalitarianism and cooperation) and supporting alternative scientific research programs. These would be research programs emphasizing the mutual co-determination of genes and their cellular environments, of cellular and physiological states of organisms and physiological states and processes of organisms, their behaviors and their environments. Just as research into biological bases of socially significant traits and the ideologies of individualism and determinism are mutually dependent, so are their alternatives. It is these the critical social movements must encourage if we are to turn our associated technologies to general, rather than particular, benefit.

Notes and References

1. Evelyn F. Keller, "Physics and the Emergence of Molecular Biology," *Journal of the History of Biology* 23 (1990): 389–409.

2. James Watson and Francis Crick, "Molecular Structure of Nucleic Acids," *Nature* 171 (25 April 1953): 737–38.

3. Many prenatal screening clinics require that clients agree to terminate a pregnancy if the fetus is discovered to carry any one of several deleterious genes, which seems to diminish rather than expand a woman's freedom.

4. The anthologies edited by Gena Corea, Patricia Spallone and D. L. Steinberg, and Michele Stanworth show the range of feminist opinion regarding the new reproductive technologies. Cf. Gena Corea, *Man-Made Women: How the New Reproductive Technologies Affect Women* (Bloomington, IN: Indiana University Press, 1987), Patricia Spallone and D. L. Steinberg, *Made to Order, The Myth of Reproductive and Genetic Progress* (New York, NY: Pergamon Press, 1987), and Michele Stanworth, *Reproductive Technologies: Gender, Motherhood, and Medicine* (Minneapolis: University of Minnesota Press, 1987).

5. Ruth Doell, "Reproductive Technology and Gene Engineering: Whose Science is This?," in Joan E. Hartman and Ellen Messer-Davidow (eds.), *(En)gendering Knowledge: Feminists in Academe* (Knoxville, TN: University of Tennessee Press, 1991).

6. Aldous Huxley, *Brave New World* (New York: Bantam Press, 1946, 1955).

7. Cf. Robyn Rowland, "Reproductive Technologies: The Final Solution to the Woman Question?," in Rita Arditti, Renate Duelli Klein, and Shelley Minden (eds.), *Test Tube Women* (London: Pandora, 1984).

8. Sally Allen and Joanna Hubbs, "Outrunning Atalanta: Feminine Destiny in Alchemical Transmutation," in Sandra Harding and Jean F. O'Barr (eds.), *Sex and Scientific Inquiry* (Chicago, IL: University of Chicago Press, 1987), p. 84.

9. Ibid., p. 90.

10. Philip Handler (ed.), *Biology and the Future of Man* (New York, NY: Oxford University Press, 1970), p. 968.

11. Cf. Mary O'Brien, *The Politics of Reproduction* (Boston, MA: Routledge & Kegan Paul, 1981).

12. "Resolution from the FINRRAGE Conference July 3–8, 1985, Vallinge, Sweden," in Spallone and Steinberg, *Made to Order*, op. cit., pp. 211–12.

13. Helen E. Longino, *Science as Social Knowledge* (Princeton, NJ: Princeton University Press, 1990).

14. Mary Hesse, *Revolutions and Reconstructions in the Philosophy of Science* (Bloomington, IN: Indiana University Press, 1980), pp. 63–110.

15. Pierre Duhem, *The Aim and Structure of Physical Theory*, trans. Philip Weiner (Princeton, NJ: Princeton University Press, 1954). See also Longino, *Science as Social Knowledge*, op. cit.

16. Daniel Koshland, "Sequences and Consequences of the Human Genome," *Science* 246 (13 October, 1989): 189.

17. One wonders what technology Koshland has in mind. Until gene therapy is actually available, the only preventative measure for diseases directly caused by genetic mutations is the abortion of fetuses carrying such genes.

18. Evelyn F. Keller, "Nature, Nurture and the Human Genome Initiative," in Daniel Kevles and Leroy Hood (eds.), *The Code of Codes: Scientific and Social Issues in the Human Genome Project* (Cambridge, MA: Harvard University Press, 1992).

19. But see Handler, op. cit., p. 926.

20. In California, mental illness became a cause of homelessness by governmental fiat in the 1960s, when certain categories of mental illness were declared not to require institutionalization and many individuals were simply released to fend for themselves. But this is surely not the kind of causal relationship Koshland has in mind.

21. Cf. Miranda Robertson, "False Start on Manic Depression," *Nature* 342 (18 Nov 1989): 222.

22. Evelyn F. Keller, Berkeley International History of Science Summer School Lectures, University of California, Berkeley, 20–24 July 1992.

23. "Gender" is used to refer to the behavioral and temperamental differences obtaining between the sexes, while "sex" refers to identification by physiological potential for reproduction.

24. The excitement over Simon Levay's finding that a set of interstitial nuclei of the anterior hypothalamus is smaller in homosexual men than in heterosexual men is testimony to the power of determinist ideologies. (Cf. Marcia Barinaga, "Is Homosexuality Biological?," *Science* 253 [30 August 1991]: 956–57.) Another somewhat amusing example is provided by the United States weekly magazine *Time* which ran a cover story in early 1992 on biology and sex differences. The cover proclaimed: "Why Are Men and Women Different? It isn't just upbringing. New studies show they are born that way." The story itself (Christine Gorman, "Sizing Up the Sexes," *Time* 139, no. 3 [20 January 1992]: 42–51) detailed research on the mutability both of behavior and of neuronal connectivity in response to environment; hardly what the cover promised.

PART VI

Eccentric Positions

THE SUBJECT OF technical action stands outside the technical apparatus which mediates between itself and the objects on which it operates. It looks in on the world from an "eccentric position," remote from the everyday lifeworld. To the extent that technical action becomes the paradigm of action in general in the course of the scientific-technical revolution of modern times, this eccentric position acquires broad social and ethical significance. It is a kind of alienation that may perhaps be natural to the human species but which takes on menacing proportions in a technological society.

In "Sade, the Mechanization of the Libertine Body, and the Crisis of Reason," Marcel Hénaff explores the consequences of this alienation carried to its logical conclusion. Once the human being is reconceptualized by Enlightenment science and philosophy as a machine, it is possible to imagine a world without morality. The result is not merely a philosophy of sexual freedom but, stranger still, an astounding anticipation of the universe of industrial capitalism. Sade's "libertines" accumulate the capital of their fantasies, assemble a workforce of victims and supervisors, construct veritable assembly lines of pleasure composed of human bodies, and keep accurate accounts of their misdeeds and triumphs like so many investors on the stock market. Rationality appears, in this universe, exclusively as a planning intelligence detached from moral community and social life. The human body itself is conceived as a mere device, decomposable into its usable parts and ultimately disposable according to the materialist doctrine of the epoch. Adorno, Horkheimer, and Foucault were naturally attracted to this monstrous work which, through an excess of rigor and consistency, exemplifies the moral limits of the Enlightenment and the civilization it shaped.

Under the influence of Heidegger, Hannah Arendt developed an influential humanist critique of modern society that also probes the alienation of technological society. Pieter Tijmes evaluates her position in his chapter, "The Archimedean Point and Eccentricity: Hannah Arendt's Philosophy of Science and Technology." For Arendt, modernity is characterized by the breakthrough to a disincarnated "Archimedean point" beyond earthbound experience. Science and technology act upon the world from this alienated position, separating human beings not so much from nature as from the social world they have themselves created, and which is their true home. Tijmes argues that the alienation Arendt attributes to science and technology is due rather to the inherent reflective capacity of the human species. In reality, the integration of science and technology to earthbound existence goes on apace, as we bring life and mechanism together in a new unity. Today we are threatened less by an essential alienation than by the lack of conscious awareness and public control of the increasingly mechanical conditions of our existence.

13

Sade, the Mechanization of the Libertine Body, and the Crisis of Reason

Marcel Hénaff
Translated by Anne-Marie Feenberg

D OES IT MAKE sense to wonder what contribution the study of fiction can make to the history of science and technology? The answer is not obvious because it assumes the prior clarification of another question, i.e., what is the relation between works of fiction and the other activities of a given culture? One wonders, for instance, what a tragedy of Sophocles, a comedy of Molière, or a novel by Dickens can tell us about Greek science, seventeenth-century physics, or nineteenth-century industry. Framed in this manner, these questions do not lead to an understanding of possible links between the available data. They make sense only if the science and technology of a given era and culture do not form separate spheres but are linked to models of representation and to conditions of development and transmission which determine a rhythm of invention and a more or less powerful integration among the activities of a society.

However, the history of science and technology has been, and largely continues to be, based on the notion of an all-conquering reason. This approach is increasingly questioned not so much for its inaccuracy as for its limitations. Transformations of knowledge and technology are inseparable from more general transformations in a society or rather in a given culture (or in a larger totality called civilization). For instance, the determining role of the phonetic alphabet in the development of Greek science is well known;[1] but one could also point to the role of Athenian democracy,[2] or of techniques of argumentation in ancient culture as a whole,[3] or also of conceptions of harmony and proportion; besides these intellectual conditions, the role of the educational institutions and the prestige of the acquisition of competencies should be mentioned.[4] With regard to an entirely different period, one could show to what extent scientific discovery was encouraged by the development of commerce in England, especially since the seventeenth century. Vice versa, when new knowledge and new technical practices begin to converge, they generate a paradigm that permeates all the representational systems, narrative models, and, more generally, the various forms of artistic production.

The idea is not to look for a reflection of infrastructural conditions in representations, or in other words to privilege one referent among the many interacting ones. It would be better to speak of a generalized space of translation, as does Michel Serres.[5] The effects of these translations in the various fields of knowledge or production of forms are neither identical nor predictable. Scholarly investigation must establish them case by case. Indeed, it may be that what makes a work specifically artistic is its ability to deal with and therefore to translate the elements of a paradigm in an unpredictable way. This is a far cry from a simple determinism, but rather belongs to the realm of what Lévi-Strauss has defined as "*bricolage*," that is, the ability of symbolic thought to construct coherent totalities out of disparate pieces, to transform fragments into crystals.[6] In short, this practice is both very much determined by its materials and very unpredictable in its syntheses.

This is the perspective from which I would like to analyze a work of fiction which constitutes an exceptional case, not only in French literature but probably in world literature as well. I am referring to the texts of the Marquis de Sade, specifically to those that are qualified as libertine. The uniqueness of this work consists in the extreme radicalism of its obscenity and its violence. Neither in its own nor in later times did it seem possible to give the reading public access to these texts. Today this prohibition has been lifted, at least in the Western world. It is acknowledged that this work must be read for what it is, that is, a work of fiction, a provocation that is symptomatic of an epoch and a civilization, not an agenda for madness or a call to murder or rape. The risk here, as Georges Bataille[7] feared, is perhaps that one will then no longer perceive its cry of revolt and its ultimate force. On the other hand, a symptomatic reading may make it possible to do something that earlier readings neglected: to try and understand the philosophical and cultural conditions of this supposedly indecipherable text.

But where to start? I think there is an obvious guiding thread, the representation of the body. It is very easy to show that Sade's entire intellectual system is organized around the image of a body which is reduced to functional mechanisms modeled on those of a machine. Unquestionably, there is a very precise link between this idea of extreme libertine behavior and the beginnings of industrial mechanization. This then also implies that Sade assimilated the lesson of Cartesian dualism as the philosophical condition underlying his model of the body.[8]

However, this approach to textual epistemology is only the first part of an investigation which should lead us to a more fundamental philosophical question: what is the significance of the fact that this representation of the body and this portrayal of extreme forms of violence appears just as Enlightenment optimism seems to celebrate its uncontested success? At any rate, those are the questions asked by Adorno and Horkheimer on the one hand, and Michel Foucault on the other. I propose to return to them here.

The Construction of the Body-Machine and the Organization of the Orgy

What profoundly distinguishes Sade's text from the numerous libertine texts of his century is the originality of the model of the body on which he consistently confers a completely mechanical status. This is why the Sadian libertine in no way resembles the great seducers of earlier times. Don Juan, for instance, was still a dreamer; he wanted to seduce, to be loved for himself; the look of the loved one could still make him tremble. In spite of everything he was still nostalgic for courtly love; he still dragged along a chivalrous spirit in the new spaces that technical and industrial reason had begun to carve out. The classical libertine does not see the change that is sweeping him along; his courtship is old-fashioned but he already has a passion for large numbers; in that respect, he lacks the means to satisfy his desires. Sade will provide them. He will invent a body for the libertine to match the fantasies generated by the new order of things. He will subject that body to a rigorous process of division and abstraction and reduce it to a mechanism of *"partes extra partes"* foreshadowing its industrial destiny. Anything that might signal an amorous relation (such as indecisive glances, the aura of desire, the ambiguous signs of the lover) disappears in the combinatory arrangement of partners, in the programmed pressure of pleasures, in the frenzied exploitation of the victims. Rather, what becomes imperative is a rigorous "project" where pleasure arises from the mechanical mastery of bodies and a strict accounting of operations.

How does Sade construct this mechanized body? We need to consider this carefully if we are to understand what is at stake philosophically. Let us remember that for the body to be conceived as a machine, it had to be thought first as simple matter defined by extension and movement according to the principles of Cartesian philosophy; furthermore, the various organs had to be understood as mere instruments according to the theses of the "iatromechanical" medicine arising from that same tradition. In short, Descartes and the Cartesians wanted to consider the organic body in the framework of physics and thus had to assign it the same status as other bodies in the natural world.[9] In order to do so they had to combine a physics of solids and a physics of fluids (which made it possible to define the status of muscles, blood, the nerves, and the humors) and to associate them with mechanics (which made it possible to deal with mobility, the circulations of flows, gestures, and various movements). This approach was established by a philosopher and doctor, La Mettrie, in the eighteenth century in a work that quickly became famous, *L'homme-machine* (1774). Relying on medical observation and going beyond even Cartesian positions, he declared: "Let us therefore conclude boldly that man is a machine and there is only one substance, variously modified, in the whole universe."[10]

It should be stressed that by excluding the hypothesis of the mind as an organizing principle, the thesis of the body-machine culminated in the idea that our organs function autonomously and are regulated and react independently of our will. The corollary of this thesis is that the mind (if it exists) is not separable from the body, or rather that it is simply a bodily expression. Sade assumes the materialist position: the body is an assembly of organs which have their own unchangeable laws.

Sade combines Cartesian mechanics and Diderot's naturalism to arrive at the concept of the autonomy of the living. Death is but the return to the inorganic state of the temporary entity made up of cells which forms the living body. Nature gives life and takes it away; it engenders forms and dissolves them. Why, asks Sade, can't man follow this example? Why would he, in doing so, be more criminal than nature itself? The torturer-libertine is only a surgeon or chemist whose hand is guided by nature. All he does is play a role in the vast movement of production, transformation, and destruction of the universe itself.

The hypothesis of the man-machine is necessary for libertine behavior which knows neither the moral law when faced with the suffering of the other nor any limit in the exploitation of the desired body. (Let us note, however, that this logic pertains to Sade and not to La Mettrie himself, who on the contrary tends toward an altogether peaceful hedonism.)

Libertine Arithmetic

Thus Sade presupposes a body understood purely as a mechanism of separate functions and humors, without interiority, without unity other than that imposed by an external will. This is how he can make the body into an instrument for orgies, represent the orgy itself as a mechanical arrangement, and transform the relations between subjects into relations of simply quantifiable bodies and organs.

Indeed, it should be noted that the quality of the utilized organs, the quantity of the mobilized bodies, and the scale of the accomplished orgies are always expressed in precise numbers. This produces four types of basic operations which could be called an arithmetic reduction: (1) measuring the organs; (2) evaluating the quantity of bodies; (3) counting the acts; (4) drawing up the balance sheet of the operations.

Measuring the organs. The description of bodily organs devoted to pleasure never goes without an evaluation of numerical quantities; this is because the very statement of those measurements is supposed to found the value of the designated object. This is the case for the male genitals. Here are some famous instances: Noirceuil: "An instrument of seven inches thick and eleven long"; Saint-Fond: "His muscular member was about seven inches long and six thick"; Claude: "Circumference of nine inches by thirteen inches long"; Minski: "eighteen inches long and sixteen thick."

The statement of these measurements aims to show, by exaggeration, that the desirable body is only an arrangement of itemizable pieces from which specific pleasures can be extracted. The variation and precision of the numbers are supposed to introduce specific and striking detail where the libertine narration tends to produce mere repetition. This demand for *detail* is explicitly formulated in the text, either by the narrator (female most of the time) who offers to satisfy it or by the public which expresses it. ("These details excite my imagination to such a degree!" [VIII, 285].) It should be pointed out that the answer always consists of numbers. Hence the surprising gap between the general character of the description of places, persons, and actions and the extreme precision of the numerical data about organs and acts of debauchery. It is as if excessive precision was supposed to compensate for the rather obvious lack of verisimilitude of the narrated actions. Obviously, this involves enormous irony since it introduces extreme realism in a domain where literary convention calls for silence and where these details concern body parts—sexual organs—which no acceptable narrative can designate except through a metaphorical ruse.

This obscene numbering is the more striking since the measuring concerns circumstances and an object which the frenzy of passion normally prefers to ignore. In *La Philosophie dans le boudoir*, for instance, the libertine Dolmance is sodomized by his valet in the presence of young Eugenie, who has been invited to discover these novelties. The following astonishing dialogue takes place on this occasion:

> *Dolmance*: . . . Oh! Christ! what a bludgeon! I have never had one like this. How many inches remain outside, Eugenie?
> *Eugenie*: Almost two!
> *Dolmance*: That means I have eleven in my ass. How delicious! (III, 455)

It is tempting to see this measuring activity as an example of fetishistic fascination. This would mean failing to understand that such selective partitioning of the body by the look belongs to the practice of taxonomic reason. In the context of the mechanization of the body, the emphasis on a specific element (this or that organ, its form, size, and performance) is but the choice of the best way to establish a connection between bodies understood as operational apparatuses. This element, which always remains the same, fits in a series of other elements that have the same function: in short, all the parts of the body that are identified as privileged objects of desire become "a class" and are repeated in the individuals who fill the narrative.

Evaluating the quantity of bodies. It would seem that what most satisfies the libertine imagination in Sade is quantity. "Nothing excites the imagination as much as large numbers." There are no doubt many reasons for this. First of all, quantity denotes luxury and abundance, and hence political and economic power. Moreover, it offers security to desire: not only must the specter of the shortage of objects of pleasure be removed, but the reserves should hardly be touched even after the greatest squandering (of energies, mobilized individuals, sacrificed victims, etc.). These mechanical models are thus inseparable from economic models, as we will see later.

It must be noted that Sade situates his orgies in places where a large number of individuals are already concentrated, such as monasteries, or which can accommodate them, such as chateaux where the basements are living storerooms for lust and crime. In every case, precise accounting insures that there is an inventory of the stock of available bodies. Here are a few examples:

In the manor of the Society of the Friends of Crime: "two seraglios are allocated for the members of the Society. . . . One consists of three hundred young boys from seven to twenty-five years old, the other of the same number of girls, from five to twenty-one years old" (VIII, 408).

At Minsky's: "I have two harems. The first one contains two hundred young girls, from five to twenty years old; there are two hundred women from twenty to thirty years old in the second one. Fifty servants of both sexes are employed to serve this considerable number of objects of lubricity, and for recruiting I have one hundred agents scattered in all the big cities of the world" (VIII, 559).

At Brisa-Testa in Constantinople: "I have seen more than a thousand individuals of both sexes during that year" (IX, 300).

It is easy to see that such a demand for quantity leaves no room for amorous relations; this is a world of interchangeable bodies in a context of complete indifference to subjective considerations.

Counting the acts. To the accounting of bodies is added that of acts, at least those that concern erotic relations reduced to sexual connections. Here too any possibility of the lyrical expression of bodies is submerged by the enormity of the numbers and the coldness of the balance sheet.

In the middle of an orgy at the house of her accomplice Clairwil, Juliette, for instance, evaluates the results by doing her calculations: "Every team of eight doubled up, changing women and ways of fucking. . . . Clairwil thus was fucked fifteen times in the mouth, ten times in the cunt and thirty-nine times in the ass: I, forty-six in the ass, eight in the mouth and ten in the cunt; in all two hundred times for each of them" (VIII, 468). Sade cannot resist the pleasure of such inventories and adds this note at the bottom of the page: "Thus, without counting the mouth which does not produce a strong enough sensation to be counted, those two honorable creatures were fucked, Clairwil one hundred and twenty-five times, and Juliette one hundred and ninety-two times, in the cunt as well as the ass" (VIII, 468). Juliette's advantage. The passage ends as follows: "We felt we had to make the addition in order to spare the women this trouble, who would certainly have interrupted their reading to do so" (VIII, 468). It was still possible to write this in the eighteenth century. . . .

The accounts drawn up for acts are often accompanied by an evaluation of the quantity of sperm that has been spilled. The king of Naples presents his stock of debauchery: "The vigor of these men equals at least the superiority of their organs; not one of them can guarantee less than fifteen or sixteen discharges; not one of them who does not lose at least ten to twelve ounces of sperm with each ejaculation; this is the elite of my kingdom" (IX, 402).

Drawing up the balance sheet of the operations. It is well known that the passion for balance sheets is one of the distinctive traits of obsessional rites. This passion is related to anality less because of the objects counted than because of the desire to control (or the obsession of loss of control) that calculations reveal. When the Sadian libertine does his accounts and wants them to be accurate (to the very unit) it is not because one more "shot" would have modified his pleasure, but because the latter emerges precisely out of this claim to accuracy, however arbitrary. Pleasure does not just arise from the accounting that is done during the action but also in the summary balance sheet that follows it. This drawing up of accounts puts the finishing touches on pleasure by making it definitive (the account is closed), glorious (the account is enormous), and, especially, verified (the account is known). Here is for instance the result announced after Juliette's visit to the monastery of Bologna: "All the novices, several nuns, fifty residents, one hundred and twenty women in all, went through our hands" (VIII, 550); and this balance sheet at the King of Naples': "We sacrificed in all eleven hundred and seventy-six victims, which amounts to one hundred and sixty-eight for everyone, among whom six hundred girls and five hundred and seventy-six boys" (IX, 412).

However, among the numerous balance sheets which abound in the Sadian text, the most astonishing, both because of its meticulousness and an expository style borrowed from actuarial forms, is undoubtedly the one which closes the narrative of *Les 120 Journées*. This surprising summary would deserve to be quoted in its entirety to prove the point; here at least is the conclusion:

This summary shows the use made of all the subjects, since there were forty-six in all, that is:

Masters	4
Old women	4
In the kitchen	6
Women historians	4
Sodomites	8
Young boys	8
Wives	4
Young girls	8
Total	**46**

Thirty of them were sacrificed and sixteen went back to Paris. Total count: Massacred before the first of March

In the first orgies	10 persons
Since the first of March	20 persons
And those who went back	16
Total	**46 people** (XIII, 429–31)

The Combinatory Model

Because the body is cut up, mechanized, counted, erotic relations are reduced to mere combinations and must be realized in the organization of a system of variations designed to establish the greatest possible number of connections between available bodies and their organs. Since pleasure is quantitative (or, more precisely, linked to the representation of quantities) the combinatory process achieves the most logical solution to the demand for profitability conveyed by the balance sheets. Three operations dominate this combinatory model: planning, variation, and saturation.

Planning. No orgy in Sade simply either starts spontaneously or is sparked by an atmosphere, a mood, or a whim, as is the case in a number of libertine narratives of the same period. In *Juliette,* and even more so in *Les 120 Journées,* one of the protagonists must first describe the actions that are to take place. The orgy is possible only if it has been carefully planned, if it follows a precise program, if it has been discussed before being performed. This means several things in Sade: first that the orgy is no anarchic pleasure but is an intellectual exercise; in other words, it can take place only if it conforms to a system of representations. What characterizes the victims is precisely the fact that they do not partake of this intellectuality. For instance, Noirceuil, Juliette's accomplice, talking about the catamites who fell asleep during a speech, exclaims: "Those creatures are imbeciles; they are machines for our lust; they are too stupid to feel anything" (VIII, 147). We really do have here the Cartesian model of a body-machine guided by an external thought and will. Second, the orgy needs to confirm an order (distribution of places, assignments of postures and acts, forecasting of the sequence of actions). Finally, the orgy is from beginning to end framed by a discourse which precedes, accompanies, and concludes it. In other words, Sade's materialism is not an empiricist sensualism but a rationalist materialism. A body without mind does not mean a world without reason. This body, one could say, never loses its head insofar as the latter is the locus where representations are organized and where the orders originate. The head is to the body what the master is to his subjects: it thinks and decides for the organs that remain subjected to it. Only through representation does pleasure arise.

Variation. In general, the outline of the program itself describes the sequence of variations that the different bodies are charged to perform. But the mechanism of the group-body can, just like a real cybernetic machine, modify or enrich its program as it unfolds either through new recruits or through the fantasies that occur to one or another libertine during the action. In fact, almost every production thus goes beyond its initial program, for it is precisely in the indefinite *supplement* of combinations that the unlimited capacity for imagined pleasures is formulated. "All three of us plunged back in the ultimate excesses of lewdness with a thousand new postures" (VIII, 360); "The situations varied seven times and seven times my semen flowed in their arms" (VIII, 360).

But the most precious advantage of this combinatory effort is *surprise* at the configurations it achieves: Libertine value is defined above all by the unforeseeable (or not-yet-seen). It is this unforeseeable element that sparks pleasure, not only because it insures the production of the differential trait which symbolizes and condenses all the differences, but also because it constitutes the paradox of producing an original event out of mere combinations: "The entire libidinous act consisted of three scenes: first, while I awakened the very sleepy activity of Mondor with my mouth, my six companions, who were gathered in three groups, had to portray under his eyes the most voluptuous postures of Sappho: *none of their postures could be the same; they had to vary at every moment. Imperceptibly the groups were mixing, and our six lesbians, exercised for several days, formed the most original and most libertine picture imaginable*" (VIII, 158–59; my emphasis).

This way of searching for original configurations by the technique of variation determines the entire narrative logic of the *120 Journées*. As we know, this narrative proposed to list and describe 600 different "passions," 150 per month for four months. Some of the multiple episodes recounted by the accredited women storytellers (called historians) differ only by a trivial element. However, this element is enough to make the difference, in other words to define an original passion. The smallest divergence becomes the basis, as qualitative difference, of the uniqueness of the configuration; by establishing a new unit, this divergence increases the total sum. Hence the following paradox: quantity is sought (so many bodies, "shots," discharges, etc.) but qualities make it possible to increase the number of additional units. Libertine desire cannot get enough variations and yields only to accumulation, but at the same time every variation must have its own irreducible singularity; this seemingly contradictory double requirement can however be met by the libertine sensibility which is capable of perceiving in the smallest divergence a pertinent difference. The reader is invited to grasp the extreme subtlety of the distinction in the introduction to *Les 120 Journées*.

> As for diversity, be sure it is right; carefully study a passion which seems to you no different from another, and you will see that there is a difference, however faint; that passion alone has the refinement and the tact that distinguish and characterize the kind of libertinage we are discussing. (XIII, 61)

This is the erotic equivalent of the Leibnizian principle of indiscernibles (in other words the principle of discernibility). Leibniz poses the problem in these terms: "if a portion of matter is in no way different from another and is the same in quantity and configuration, and if the state of the body at a given moment is no different from the state of that same body at another moment except by the transposition of portions of matter, equal in quantity and configuration and similar in every respect, then it clearly follows from the perpetual substitution of indiscernible portions, that there is no way to distinguish the states of the corporal world at different times" (*De ipsa Natura*, §13).

This is why, according to Leibniz, we must posit the hypothesis of a qualitative difference that founds the singularity of each substance and affirm "that there is never a perfect similitude, which is one of the important axioms I have discovered" and also "that it is not true that two substances resemble each other and are different *solo numero*" (*Discours de la metaphysique* §9). It would be enough to substitute the term *passion* for the term *substance* to have a perfectly Sadian phrase.

Saturation. We know that the definition of every combinatory includes the notion of saturability (as with an axiomatic). This is evident in the Sadian combinatory. Saturation is to the libertine body what depth is to the lyrical body: its utmost completion. What does this operation look like? Several figures are possible, including either the saturation of the sexually excited body or of the predicates of its action.

Saturating the body means first of all sexually occupying all its spaces: "Desiring that not a single part of my body remain empty" (VIII, 139); "no genitals escaped my hands, no part of my body remained unsoiled" (VIII, 425). "When we come to be fucked, we want it in every part of our body" (VIII, 437). But saturating the body also means connecting it to a maximum number of other bodies: it means realizing the body-group. If it is true that saturation is the very form of libertine achievement and its ultimate fulfillment, it is easy to understand why in Sade situations involving couples are rare and always temporary. The norm here is the group, the orgy; the couple relationship appears only during a first meeting, or to establish a pact, or to share a secret, or at the unspeakable intimate meeting of the torturer and his victim in the secret chamber.

At a second level saturation concerns the predicates of action: in fact it concerns a purely linguistic pleasure consisting in either the accumulation of different predicates related to the same subject or the permutation of the same predicate between different subjects. In the first case we deal with situations such as the following: "Here I am, satisfied. . . . I am being fucked, I fuck a virgin, I have my wife sodomized, nothing is lacking any more for my ardent pleasures" (VIII, 138); "I was covered with maledictions, curses, *I committed parricide, incest, I murdered, prostituted, sodomized! O Juliette, I have never been so happy in my life!*" (VIII, 254). Roland Barthes gives an excellent analysis of the pleasure of accumulation: "The counting up of pleasures provides an additional pleasure which is that of the addition itself"; "This superior pleasure which is purely formal since it is after all a mathematical idea, is a linguistic pleasure: that of deploying a criminal act under different names."[11]

The reverse of this pleasure is the permutation of the same element between different subjects. At Silling, for instance, the four libertines conduct wedding ceremonies by successively marrying their girls, their sodomites, their catamites, and then force them to marry each other. Thus applied, the predicate marriage is completely blurred, as is that of "sexual identity" in this experience: "That evening, the quadrilles changed sexual identity: all the little girls changed into sailors and the little boys into *grisettes.* It looked delightful; nothing excites lust as much as this little voluptuous exchange" (VIII, 138).

These attempts at combinatory saturation (there are so many examples of them) confirm for us, if need be, that only the completion of the inventory interests the Sadian libertine and that only the journey through the passions can satisfy an expectation which is no longer psychological but encyclopedic.

To Say Everything, to See Everything: The Erotic Panopticon

"Philosophy must say everything," proclaims Juliette (IX, 586). Sade adds to the words of his character in the preface of *Les 120 Journées*: "If we had not said everything, analyzed everything, how could you expect us to have guessed what suits you?" (XIII, 61). This ambition to have both totality and excess, which tries to attain the whole through what exceeds it, defines the libertine discourse as the encyclopedia of obscenity, that is to say as the system of total exposure. Speech has to be uninhibited and the voice of the master legislates over the entire space encompassed by his vision. Nothing can escape such a vision precisely because, by hypothesis, everything has been brought into the picture, which coincides with the system (let us thus designate the totality of the discourse intended by the libertine). What is sayable has to be visible, and vice versa: "Everything must be seen" (III, 387). This is saturation without remainder; the possibility of an uncontrolled supplement is averted by the hypothesis of the "secret chamber": that which cannot be seen can still be designated as taking place in the same sort of spot where everything else takes place, at the limits of what is imaginable and kept in reserve for a future telling: "It has always been impossible for me to find out what was happening in those infernal chambers" (XIII, 297).

The absence of any possible retreat outside the picture, the demand for a total exposure, is what constitutes *obscenity*: the space of the all-seeing master. Hence the request for complete and immediate nudity. Sade does not fool around with halftones and indirect suggestions; his mechanistic model is too radical for that. It is also because the Sadian libertine is all-seeing that Sade has no particular interest in voyeurism. It can appear among the numerous passions listed in *Les 120 Journées*, but that is all. Because in principle everything is visible in Sade, voyeurism is not an interesting perversion; this indeed presupposes a ban on looking, which of course is excluded in the Sadian world. Everything has to be exposed to view, without mediation, without resistance, without delay; just like the body, the picture must be uncovered at once. Between libertines, to see means to be seen immediately. There is a sort of integral vision, obtained by the unceasing shadowless interaction of all the lines of sight: "Groups were arranged so that everyone could enjoy seeing the pleasures of the others"; "No more private exchanges, no more tête-à-tête, said the King: we must now operate in each other's view" (IX, 408); "The arrangement of the shrubs was so regular that there was no table from which any of the others would be unseen; and as a result of the cynicism which had organized all this, except for those in the salon, the lewd acts at the supper could not escape the observant eye" (VIII, 424).

This arrangement becomes a kind of *erotic panopticon* where nothing offered by the view of the bodies remains hidden. "Fucking, fucked, looking at fucking" (VIII, 138). The exposure is total. But unlike the *panopticon* of Bentham (so well analyzed by Michel Foucault),[12] which aims at the non-reciprocal omnipresence of police surveillance—seeing without being seen—in becoming erotic Sade's panopticon also becomes playful, reciprocal, and multicentered: instead of one single locus of vision there are as many as there are looks, that is, as there are connected bodies.

With its ceaseless permutations, the group of libertines occupies the totality of possible points of view so far as the pleasure of seeing is concerned. Of course, they have only a limited point of view at each moment, but at the end of the orgy the variations of their positions or their roles will have allowed them to view the picture from every angle, with all its figures and all its movements. To be sure, there is nothing like the Leibnizian God, the absolute center of vision, the point of intersection of all points of view, but the group itself constitutes a kind of *unique body* with multiple eyes, hands, and sexual organs: a completely saturated body, but a body without subjectivity (because, for Sade, pleasure, like death, remains the absolute mark of individuation). In this group-body, this machine of anatomical connections, there is no room for the look of love, no opportunity for the subjective. The circularity of the spectacle in no way implies the reciprocal recognition of minds in the transitivity of the looks. One sees the bodies and their parts, but one does not view the look itself in the sense in which it transmits a message of interiority, translates hidden feelings, and calls for a hermeneutic of expression. Here nothing is hidden nor should it be: all desire can be reduced to a countable passion. Everything can and should be laid out, broken up in parts, combined. The eyes, like the other parts of the body, are included in a taxonomy which signals them in a purely anatomical way or according to a general erotic value: "black," "interesting," "expressive," etc. (even when they are called expressive they are never seen "expressing").

This flattening out and mechanization are particularly evident in the use of *the mirror* in Sade. There is a long tradition linking the mirror to the double, the beyond, the revelation of the psyche, the threat of death. The mirror doubles space and bodies in an illusory depth and gives everyone the ability to verify or contemplate his own face; all this makes the mirror into an instrument that is related to interiority and its uncertainties. Baroque poetry in the beginning of the seventeenth century linked it to water and made of it the very symbol of an unstable evanescent subject, anxiously questioning his identity, lost in his images.[13]

Sade gets rid of this dreamy and floating specularity. He does not abandon the mirror, the classic accessory to debauchery, but he exorcises it, demystifies it, and throws out its ghosts. In short, he instrumentalizes it and brings it back into the rank of objects that are available and efficient in order to produce determined and countable erotic effects. This is perfectly clear in the passage in *La Philosophie dans le Boudoir* where the young Eugenie, initiated by Mme de Saint-Ange, hears the latter explain to her the layout of her accomplices' place:

Eugenie: Oh God, what a delicious nest! But why all these mirrors?
Mme de Saint-Ange: This is to multiply to infinity, by repeating postures in

a thousand different directions, the same pleasures for those who are experiencing them on this sofa. No part of this or that body can thus be hidden: everything must be visible; they are so many groups assembled around those chained together by love, so many imitators of their pleasures, so many delicious scenes which intoxicate their lust and which soon contribute to its satisfaction. (III, 387)

It would be hard to find a better theory of the mirror-machine, or a better way of saying that what is expected is not narcissistic contemplation but the mastery of the totality of the visible as well as the multiplication of its effects, and therefore those postures that constitute the variable details (to use the Leibnizian terminology) and increase the combinatory possibilities. It must be noted that we are not dealing with one mirror but with mirrors, in other words a plural arrangement. This arrangement does not aim at producing images where looks interrogate each other and slide into depths that make them reel, but at producing scenes on the same level as those that are offered to the view of the actors on the stage; the function of the interplay of mirrors is simply to provide perspectives that allow for omnivisibility: "Everything has to be visible" says the main character of the *Boudoir*, Mme de Saint-Ange. That is all. This is how a purely mechanistic order is affirmed. It is clear that Sade does not deal with the imaginary but with the imaginable, that is to say not with fantasies but with calculations, not with dreams but with programs.

The Factory of Pleasures

Although the last third of the eighteenth century in France is not yet characterized by industrial production, it is already the age of the factory and collective work. It is noteworthy that in Sade the settings for debauchery simulate (or anticipate in other respects) the organization of industrial work. All the necessary phases for implementing production are there: capital formation, the acquisition of raw material, the recruitment of the workforce, training of the latter, the organization of productive activity, assessment of results. All these procedures are explicitly and rigorously outlined in the organization of debauchery.

1. *Capital formation.* The four libertines of *Les 120 Journées*, for instance, gather a considerable sum of money to provide for four months of debauchery in Silling. Incidentally, we are told that this is an ongoing practice with them, as they are organized in a Society of Friends of Crime. "The society had a common purse administered by one of them in turn every six months; but the funds in this purse, destined solely for pleasures, were immense. Their excessive fortune made very strange things possible, and the reader should not be surprised to hear that two millions were allocated for the pleasures of the table and lewdness alone" (XIII, 4).

2. *The acquisition of raw material* consists of the work of the recruiters which precedes the operations in Silling. This is also true of all of Juliette's orgies. Upon his

arrival in Naples, the supervisor of the King's pleasures recounts: "I have twelve suppliers in the field and through my good offices, you will be presented every morning with twenty-four handsome boys eighteen to twenty-five years old" (IX, 344).

This "human material" (to use Marx's term) is free for the libertine. Wealth or power ensures that the poor can provide him with unlimited supplies in the form of young boys or girls that are kidnapped or bought. One is reminded here of the advertisement that Marx found in a report by a factory inspector in England in the middle of the nineteenth century: "Wanted: twelve to fifteen young boys, no younger than what can pass for sixteen years old" (*Capital*, I, chapter 15). In Sade's fiction, capital's vampirism, denounced by Marx, is already in place in the form of sexual exploitation.

3. *The workforce* consists in the mass of servants (valets, coachmen, gardeners, cooks, maids, etc.) who often take part in the orgies.

4. *The training personnel* consists in the recruiters, the pimps, the go-betweens, the guards, in short an intermediary population that is often well treated by the libertines and escapes the massacres. In a way they are the "engineers" of the debauchery. They benefit from the superior status of those who are associated with enterprise, capital, and the implementation of the libertine project.

5. *The proletariat* of the libertine enterprise, as we will call the mass of victims, are destined in the orgies to contribute to the masters' pleasure. This mass forms a body of victims at the service of the libertine, a body which one could call proletarianized because it is totally dependent, manipulated, exploited, molested, exhausted, and finally rejected after use, namely when its erotic productivity has become nil.

One could also show how the model of the factory is confirmed by the whole form of the organization of discipline and productivity as in Silling or in other places of detention of the victims: control over time, gestures, and relations, a strict hierarchy, a system of sanctions. (However, these details also belong to other models such as the monastic model, the military model, as well as to manufactures and factories, not just at their beginnings but until recent times.)

This is a quick review, confirmed by a multitude of Sade's texts, of the framework of this factory of pleasures. We still need to see how, within this framework, the operation of pleasure is achieved, how it is somehow a product produced by the mechanistic functioning of bodies. Indeed, specifically Sadian libertine pleasure is always linked to a system which through the mechanisms of labor and through series of regulated operations aims to bring about a quantified production of acts of debauchery. Humor arises precisely from the encounter between this production imperative and a goal which supposedly is its negation, as in descriptions such as the following:

Francaville offered us the most beautiful ass in the world: two children placed next to this ass had to open it, wipe it and direct to the hole the monstrous organs dozens of which rushed into the sanctuary; twelve other children removed the waste. I have never seen such an agile performance. Those beautiful organs, thus prepared, went through many hands to those of the children who were to introduce them; they disappeared in the ass of the patient; as soon as they left, they were replaced by others; all this was done with inconceivable nimbleness and promptitude. (IX, 366–67)

Elsewhere, Sade describes an even more complicated operational scheme where everyone is assigned a specific task which must be performed with precision and speed so that the totality of the bodies functions as an ordered and efficient machine, as a rotary press functioning at full speed. "Fifteen girls arrive three by three; one whips, another sucks, and the other shits; then the one who shits, whips, the one who sucks, shits and the one who whips, sucks. So he uses all fifteen of them. . . . He starts this game over again six times a week. It must go very quickly" (XIII, 359). Sade always insists on the speed of execution in keeping with the specialization of the gestures, i.e., the very definition of productivity which, according to Marx again, consists in "producing an ever increasing quantity of values in ever decreasing amounts of time" (I, chapter 14).

The existence of a model of mechanical fabrication is undeniable here. However, one should not stretch interpretation too far. The way the model plays with the referent is flexible and complex: sometimes we are dealing with a type of workshop activity, at other times with what is called heterogeneous manufacture or serial manufacture. It is obvious that the dominant paradigm in Sade is a world where machines begin to impose themselves and where mechanical functioning is grasped and reformulated as a way of desubjectivizing bodies, as the affirmation and production of the pleasure of quantities. In that context, neither the lyrical impulse nor the idea of the soul can hold out. This functioning gives Sade's materialism its particular strength, sanctions it, and leaves no room whatsoever for any religious or even moral vision. The love relation is reduced to a simple muscular movement and to reactions of the nerves and humors.

However, we should not forget that we are dealing with a work of fiction: its function is neither to prove nor to inform, judge, legislate, or make believable but at most to show and experiment. Models from other domains are taken up and translated and at the same time reworked: such is the role, both playful and critical, of a work of fiction. One cannot reduce it to a historical and sociological document without stripping away its very strength. Of course the work of fiction contains those dimensions but they are extras. The positivistic approach would only flatten rather than illuminate it. This text, no matter how rigorously it reproduces the history amid which it is written, responds to it primarily according to a different logic, that of "mimesis" which shifts its forms, renders them indistinct, parodies them, plays with them, disfigures them sometimes, and thereby makes them explicit. Such is the power of a work: while it grips us it also keeps us at a distance from what it presents. It emerges in the middle of history as a question asked of that history. In the case of Sade's work, this question is linked to the very outrageousness of what is represented. Libertinage does not have the familiar form of mere eroticism or parlor games. The reduction of bodies to machines, the violence visited upon them, the surprisingly methodical character of the activities are evidence of something which we must now face and discuss.

Sade and the Crisis of Reason

It should be possible to prove that there is a link in Sade's work between a certain model of the body and a certain stage in the development of technology and industry. But an argument satisfying our epistemological curiosity does not address another properly philosophical question which could be formulated as follows: what does the appearance in our history of such a strange work as Sade's mean? The mechanization of bodies, the madness of the orgies, the representation of incredible violence done to bodies, to what do they refer? Are we dealing with a work that is simply monstrous, a sort of totally irrational "freak of nature," or are these frenzies and this violence signs of a crisis in the history of reason, the experience of a limit specific to Western civilization insofar as science and industrial development have become its distinctive traits in a way they have not for any other civilization?

"Sade," writes Michel Foucault, "arrives at the end of classical discourse and thought. He reigns exactly at their limit."[14] We can concur with this judgment if by "classical thought" is understood not only a program of reason as clarification and classification but also a Cartesian strategy of mastery: "To become masters and possessors of nature." This nature, dominated both by knowledge and technology, will then also be dominated by industrial activity.

Up until the very end of the classical period, reason was essentially understood as wisdom capable of modulating and articulating the types of knowledge and integrating them into the ordinary human experience of the world and of others. Henceforth, it will be defined as "mathesis" and even as "mathesis universalis" (counting) and at the same time as technique (domination, profitability). But Sade shows us that it can just as well mean, in its new form, exploitation and destruction. Sade's text appears as the enactment of the generation of a series of moments which are a process of power, pleasure, and death; they can be read in the models of the figures and the operations of the libertine body. Cutting up, dividing up, displaying, connecting the bodies and organs, programming their postures and variations, methodically extracting every possible pleasure to enjoy them finally in torture and death: all this is in Sade the development of one and the same logic. Sade shows how the body has met its technical-industrial destiny, that the banal possibility of a post-amorous relationship has now opened up and that henceforth there exist techniques of pleasure that have no relation to either seductive desire or the rites of recognition.

However, to talk about a rational program is paradoxical. To be sure, insofar as the libertine body is programmed, dissected, exposed without secret or inside, it is handed over to the indifferent examination of classifying reason. But the libertine body is a desiring body: the locus of the expression of passions and the instrument of their realization. And, passions are what classical reason finds hardest to conceptualize. Descartes has shown this in exemplary fashion; he succeeded in formulating the *Cogito* and extracting from it the certitude of the existence of God and the truth of mathematics without even having to presuppose the existence of the body and its effects. He will need to develop a considerable amount of argumentation (the entire second half of the *Meditations*) in order to reintegrate the body in the circle of the thinkable. He can do so only at a very high price since he needs no less than the hypothesis of divine intervention to account for the effects of the passions on the soul. By itself, Cartesian reason lacks the means to think such an explanation.

Without in any way abandoning the aims of reason, on the contrary, Sadian libertine thought aspires to succeed where Cartesianism failed. It wants to include on the agenda of reason that which by definition would seem to elude it. Indeed, confronted with passion, reason had lacked daring, and merely posited passion as its radical other, its limit. Reason's procedure then consisted in assigning a place to this alien fact to contain it within its limits, to give it tasks according to an external regulatory model. That model is homologous to the method of enlightened despotism that becomes predominant in the entire political and moral thought of the sixteenth and seventeenth centuries: just as the prince subjugates and pacifies an unsettled and unpredictable crowd, so reason orders and directs blind and restive passions.

In fact, reason continues to be thought of as *substance*, confronted with another substance considered as heterogeneous. Sade, in accord with an essential intuition of the Enlightenment, makes it clear that reason is a *form*, a universal form that can impress itself on every substance. There is therefore no longer any risk of passion escaping from reason. Passion appears as rebellious matter or uncontrollable energy, but reason proves that it can tame passion by subjecting it to the rigor of its procedures. Passion is a force, and can therefore be analyzed methodically like any other force, that is, it can be broken down, calculated, applied, and finally controlled.

This is the method of *libertine apathy* which consists in channeling and analyzing energy without ever giving in to enthusiasm or frenzy. For Juliette, for instance, any passionate fervor is proof of weakness which must be sanctioned by death (this is why she ends up sacrificing her companion Olympe and even her accomplice Clairwil). Libertinage requires that "passions" be classified, calculated, and programmed. This takes a cool head. In short, *the form of reason* thus penetrates *the matter of passion*, just as in physics energy enters into the calculation of a dynamic. Pleasure for the Sadian libertine does not mean a party, or sensuality or happiness; it arises only out of the certainty of success of the accomplished program. Thanks to the action of apathy, passions are purified, purged, assigned to specific objectives in an organization of countable pleasures. In short, they become operational.

This is the formalism of reason highlighted by authors such as Adorno and Horkheimer in their famous essay[15] which shows the affinity of the methodological presuppositions of Kant and Sade. Indeed, Kant's doctrine posits the autonomy of reason as its fundamental requisite. This autonomy must be understood (at least in the *Critique of Pure Reason*) as reason's power to impose its categories on data which by themselves are devoid of intelligibility. This function that is proper to reason can be recognized in the work of cognition which can operate on any matter. This formal neutrality, according to the two authors, defines a rationality that is independent of the object: "Reason is—by virtue, too, of its very formality—at the service of any natural interest. Thinking becomes an organic medium pure and simple, and reverts to nature"; "Reason is the organ of calculation, of planning; it is neutral as to goals, and its element is coordination"; "Reason is, for the Enlightenment, the chemical agent which absorbs the individual substance of things and volatilizes them in the mere autonomy of reason itself."[16]

What is the relation between this rationalism and Sadian libertinage? This is how Adorno and Horkheimer understand it:

> What Kant grounded transcendentally, the affinity of knowledge and planning, which impressed the stamp of inescapable expediency on every aspect of bourgeois existence that was wholly rationalized, even in its every breathing space, Sade realized empirically more than a century before sport was conceived. The teams of modern sport, whose interaction is so precisely regulated that no member has any doubt about his role, and which provide a reserve for every player, have their exact counterpart in the sexual teams of *Juliette*, which employ every moment usefully, neglect no human orifice, and carry out every function.[17]

They establish the relation between, on the one hand, the appearance of modern forms of sport as a rational treatment of the possibilities of the body, as an unlimited pursuit of performance, as a way to make energies profitable, and on the other hand the Sadian orgy as anticipating a maximum of itemizable pleasures in the search for a kind of sexual efficiency. This relation constitutes for Adorno and Horkheimer a very interesting hypothesis which they do not hesitate to extend to all social activities and to the system of philosophical representations. They state: "The architectonic structure of the Kantian system, like the gymnastic pyramids of Sade's orgies and schematized principles of the early bourgeois freemasonry—which has its cynical mirror-image in the strict regimentation of the libertine society of *Les 120 Journées*—reveals an organization of life as a whole which is deprived of any substantial goal";[18] "Juliette favors system and consequence. She is a proficient manipulator of the organ of rational thought."[19]

However, at this point I must raise some questions. To be sure, as this chapter has tried to demonstrate, all the organizational and methodical models implied in the Sadian orgies reflect the predominance of a rationality that is liberated from any finality and capable of imposing itself as pure technical efficiency. It is questionable however whether its historical affirmation can be attributed to Kant's thought. To reduce the latter to the establishment of the autonomy of reason and to the triumph of a formalism that serves the interests of the bourgeoisie is a risky operation and far too simplistic.

Kant does indeed affirm the autonomy of reason, but this is to emphasize that reason finds in and by itself (and not in some external theological or political authority) the foundation for the knowledge of phenomena on the one hand and of moral decision on the other. Kant summarized this project in "Was ist Aufklärung," where the first sentences already state: "Enlightenment is man's release from his self-incurred tutelage. Tutelage is man's inability to make use of his understanding without direction from another. Self-incurred is this tutelage when its cause lies not in lack of reason but in lack of resolution and courage to use it without direction from another. *Sapere aude!* 'Have the courage to use your *own* reason!'—that is the motto of enlightenment."[20]

It should be noted that Kant here speaks of *understanding* that is of reason insofar as it is interested in knowledge, an interest which is neither the only nor the most important one. In fact, in opposition to empiricism, Kant refuses to reduce reason to the faculty of organizing the means to obtain an end; this would locate man in the mere continuity of the natural world. Kant constantly affirms that reason has its own ends which are in no way inscribed in the order of nature where instinct would be more successful. Reason is the sole judge of those ends: such is its autonomy. It discovers in itself strictly speculative purposes and practical, that is, moral, ends.

Then what does formalism mean? It is the requirement that reason retains its autonomy only insofar as it can claim complete initiative in the activity of knowledge, which means in turn that phenomena are knowable only according to the a priori categories of understanding. This is the transcendental method; this is what defines it as rational. To be sure, this knowledge is independent of experience, but it can apply only to the data of experience. To say that it concerns phenomena is to say that it does not have an opinion on the nature of things in themselves or noumena. On the other hand, in its moral interest, reason legislates directly to beings endowed with free will (where it reaches the noumena): it proclaims the law as pure form, that is, as a maxim of universal legislation. This is not a question of knowledge but of something to be achieved. This order of free causality defines reason in its plenitude, that is, according to its superior finality. Kant remembers Rousseau's lesson here: reason can be realized only in a moral order which is not dependent on speculative knowledge but on recognition of the supra-natural order of free beings.

This brief reminder forces us to consider quite exaggerated any attempt to liken Kantian thought to a formalism which would reduce reason to a neutral instrument subjecting all data to operational procedures. Kant is thus probably not the best reference to explain Sadian procedures. On the other hand it is obvious that the process pinpointed by Adorno and Horkheimer very much belongs to the Enlightenment, but its sources should be sought in the legacy of Cartesianism (in La Mettrie but also that in D'Holbach or Diderot for instance) rather than in Kant (except to caricature him).

Rousseau already denounced the Enlightenment movement as the worrisome underside of civilization; it is no other than the movement of scientific and technical modernization which drastically changes not only all knowledge and the arts but also social organization, political systems, and moral attitudes. It is the very movement that Max Weber has described as the disenchantment of the world and which Benjamin understood as the "loss of aura." The remarks of Adorno and Horkheimer belong to the same type of approach: "Even injustice, hatred, and destruction are regulated, automatic procedures, since the formalization of reason has caused all goals to lose, as delusion, any claims to necessity and objectivity. Magic is transferred to mere activity, to means—in short, to industry. The formalization of reason is only the intellectual expression of mechanized production. The means is fetishized, and absorbs pleasure."[21]

According to these authors, the Sadian text exhibits this positivist reductionism; this is what we find in the functionalization of sexuality which is a distinctive feature of the modern experience: "In pleasure men disavow thought and escape civilization."[22] "What is true in all this is the insight into the dissociation of love, the work of progress. This dissociation, which mechanizes pleasure and distorts longing into deceit, attacks the core of love."[23]

This is indeed what Sade's text presages and what Foucault, less dramatically, sensed in finding there the limit of classical thought, that is that all experiential data are immediately subjected to a discursive double in a general system of representation: "This unclassifiable work shows the precarious equilibrium between the law without law of desire and the meticulous ruling of discursive representation. . . . Undoubtedly, this is the principle of that 'libertinage' which was certainly the last in the Western world (after that comes the age of sexuality): a libertine is someone who gives in to all the fantasies of desire and all its furors and in the process can, but also must bring to light their slightest movement by lucid representation voluntarily elaborated."[24]

But we must say that Sade is at the intersection of two worlds, the traditional one of aristocratic values and the operational world of technical efficiency. It is the appropriation of the one by the other that reveals the crisis of our civilization.

Conclusion

One could ask whether such a demented text as Sade's was necessary to understand the collusion between reason and domination. Or should we rather ask whether Sade is really as alien, as atypical, as he is presumed to be. Possibly not; perhaps with his excessiveness, Sade only concludes the demonstration that was begun long before him. He fills in what had been a dotted line. He extracts the last corollaries from theorems proven long ago. He makes obvious the continuity between shattering the organic body, systematically dividing it up, and quantifying it, on the one hand, and methodically rendering it profitable and subjecting it to industrial exploitation on the other. The libertine body in Sade exists only in the context of the hypothesis of mechanical reduction and unlimited submission. The eroticism of accounting and combinations implies a passion for capital returns and the abstract temporality of programming. It is easy to guess that in this manufacture of pleasure everything can be represented and accomplished, everything except love.

Without symptom, without sleep, dream, secret, or memory, the body generated by the factory of the Sadian text is the extreme product of our extreme West. This body looks like the interplay of cogs under a glass cover; the eye without eyelid tires itself out in a vision without limits, in a space without hiding place, in a light without shadows. The body is the product of a reason which since the Renaissance no longer tolerates any limit to its conquests or any obscure provinces in its empire. Saying everything and seeing everything are extended into explaining everything and develop into a grasping everything and end in enjoying everything. Pleasure appears as the last stage of a process of domination, as saturation by excess of the entire field of what can be experienced. The unlimited crime demanded by the Sadian libertine is in his eyes no more than the proof that nothing shall escape his domination. Reason is supposed to extend its range of action into madness itself. Precisely in order to do so, however, reason needed to change its status and cease to be generally understood as judgment and wisdom. It needed to be defined as an instrument and made into an autonomous set of formal procedures capable of conferring the seal of science on any activity that would proceed according to norms of efficiency and order. In short, reason can then be defined as functionality independent of any consideration of ends.

That is why the delirium of the Sadian text is not the consecration or the climax of transgression that it is generally assumed to be. The text only pushes to the extreme, to an unbearable degree, a perfectly ordinary possibility created long ago by the rationalist project of introducing instrumental efficiency into every human activity. Sade adds the domain of the most violent perversions with the paradoxical result that they are integrated in the field of reason under the cover of the operational neutrality of a method. This is why Sade touches us and upsets us. He has succeeded in carrying to the extreme, through the provocation of a text of fiction, the premonition of a new complicity—a possible but not a necessary one—between reason and domination. All the horrors of our century testify to this complicity, since this century, as no other, has been able to kill and destroy methodically, scientifically, and industrially. We are obliged to admit this today. This is why we are able to hear the warning which this strange voice has been giving us since the end of the Enlightenment; we are beginning to decipher the message of this intolerable text. What appears intolerable in it shows us what we must take care never to tolerate. The work's madness, as any madness, speaks a difficult truth that it is incumbent upon us to confront. I do not see any other way or any other reason to read Sade.

Notes and References

Bibliography of Sade's Work

The quotes from Sade's texts are taken from the French reference edition in sixteen volumes (Paris: Editions du Livre Précieux, 1966–1967); the translation of the quotes was done specifically for this study. The texts mentioned here are *La Philosophie dans le Boudoir*, vol. 3; *La Nouvelle Justine suivie de l'Histoire de Juliette*, vols. 6, 8, and 9; *Les 120 Journées de Sodome*, vol. 13.

1. See for instance I. J. Gelb, *A Study of Writing* (Chicago: University of Chicago Press, 1952); E. Havelock, *The Literate Revolution in Greece and Its Cultural Consequences* (Princeton: Princeton University Press, 1982); G. E. R. Lloyd, *Magic, Reason and Experience: Studies in the Origin and Development of Greek Science* (Cambridge: Cambridge University Press, 1979).

2. J. P. Vernant, *Myth and Society in Ancient Greece*, 2d ed. (New York: Zone Books, 1988); J. P. Vernant, *Myth and Thought among the Greeks* (London: Routledge and Kegan Paul, 1983); M. Détienne, *Les Maîtres de vérité dans la Grèce ancienne* (Paris: Maspero, 1979).

3. Lloyd, op. cit., chapter 2.

4. H. I. Marrou, *A History of Education in Antiquity* (New York: Sheed and Ward, 1956), trans. by G. Lamb from *Histoire de l'éducation dans l'Antiquité* (Paris: Seuil, 1948).

5. Michel Serres, *Hermes III, La Traduction* (Paris: Minuit, 1974). On the question of the relation between literature and science, one should also consult the stimulating essay by Marjorie H. Nicholson, *The Breaking of the Circle: Studies on the Effect of the "New Science" upon Seventeenth-Century Poetry*, revised ed. (New York: Columbia University Press, 1965). See also M. Pierssens, *Les savoirs à l'oeuvre, essais d'épistémocritique* (Lille: Presses Universitaires de Lille, 1990); V. Nemoianu, *A Theory of the Secondary: Literature, Progress and Reaction* (Baltimore: Johns Hopkins University Press, 1989).

6. C. Lévi-Strauss, *The Savage Mind* (Chicago: University of Chicago Press, 1966), chapter 1.

7. G. Bataille, "De Sade's Sovereign Man" and "De Sade and the Normal Man" in *Death and Sensuality: A Study of Eroticism and the Taboo* (New York: Walker, 1962). See also G. Bataille, *La Littérature et le mal* (Paris: Gallimard, 1973).

8. It would be difficult to evoke in the narrow confines of this study all the data of the contemporary technical and economic framework of Sade's work, especially as it relates to the development of mechanization in the second half of the eighteenth century in France. In this respect, *l'Encyclopédie* by Diderot and D'Alembert shows a remarkable view of contemporary technology as well as the impact of that technology on a still classical culture whose themes do not reflect it but whose representations are profoundly marked by it. For further documentation on the scientific and technical framework, consult J. Roger, *Les Sciences de la vie dans la pensée française du XVIIIème siècle*, 2d ed. (Paris: Colin, 1971); F. Duchesneau, *La physiologie des Lumières. Empirisme, models et théories* (The Hague: Nijhoff, 1982); G. Ganguilhem, *La formation du concept de reflexe aux XVIIe et XVIIIe siècles* (Paris: PUF, 1955).

9. For further readings on the problems of Cartesian mechanism in Descartes, consult R. B. Carter, *Descartes' Medical Philosophy: The Organic Solution to the Mind-Body Problem* (Baltimore: Johns Hopkins University Press, 1983); G. Pflug, "Descartes und das Mekanische Menschenbild," *Medizinhistorisches Journal* 17, no. 3–19 (1982); G. Rodis-Lewis, "Limitations of the Mechanical Model in the Cartesian Conception of the Organism," in *Descartes: Critical and Interpretative Essays*, ed. M. Hooker (Baltimore: Johns Hopkins University Press, 1978), pp. 152–70; R. S. Westfall, *The Construction of Modern Science: Mechanism and Mechanics* (New York: Wiley, 1971); L. C. Rosenfield, *From Beast-Machine to Man-Machine: Animal Soul in French Letters from Descartes to La Mettrie* (New York: Oxford University Press, 1941, rpt. 1968).

10. La Mettrie, *L'Homme-machine*, ed. Olms, reprint, p. 355.

11. Roland Barthes, *Sade, Fourier, Loyola* (Paris: Seuil, 1974).

12. Michel Foucault, *Discipline and Punish* (New York: Pantheon, 1977), pp. 195 ff.

13. G. Genette, *Figures of Literary Discourse* (New York: Columbia University Press, 1982).

14. Michel Foucault, *Les Mots et les choses* (Paris: Gallimard, 1967), p. 224.

15. T. Adorno and M. Horkheimer, "Juliette or Enlightenment and Morality," in *Dialectic of Enlightenment* (New York: Herder and Herder, 1972).

16. Ibid., p. 89.

17. Ibid., p. 88.

18. Ibid.

19. Ibid., p. 95.

20. Kant, "Was ist Aufklärung," in *On History* (Indianapolis: Bobbs-Merrill, 1963).

21. Adorno and Horkheimer, op. cit., p. 104.

22. Ibid., p. 105.

23. Ibid., p. 109.

24. Michel Foucault, op. cit., p. 222.

14

The Archimedean Point and Eccentricity

Hannah Arendt's Philosophy of Science and Technology

Pieter Tijmes

Different Perspectives on Hannah Arendt's Philosophy

IT WILL NOT provoke any protest to say that the work of Hannah Arendt is rich in perspectives. The interesting books and articles dedicated to her bring these perspectives into the open. Being totally different in character, they may nevertheless articulate points of special interest. For example, in *Hannah Arendt, Philosophy of Natality*, Patricia Bowen rejuvenates Hannah Arendt's philosophy from the category of natality and thus shows how important the inspiration of Augustine's words "initium ergo ut esset, creatus est homo" have been for every level of Hannah Arendt's thinking. In *Visible Spaces*, Dag Barnouw, as a second example, describes Hannah Arendt's German-Jewish experiences as a guideline for considering her work as a whole. No doubt there are more creative interpretations of her philosophy, but considered from whatever perspective, it is always as a political philosopher that she appears, with *The Human Condition* as her *opus magnum.*[1]

By calling her a philosopher one overlooks her own opinion as she airs it in her introduction to the first part of *The Life of the Mind*, where she writes: "I have neither claim nor ambition to be a 'philosopher' or to be numbered among what Kant called not without irony 'Denker vom Gewerbe.' " She considers herself a theorist of political thinking and prefers in this way to distance herself from political philosophy, which from Plato on has been impregnated with an attitude inimical to politics. This quotation is not an isolated remark but expresses a deeply rooted conviction. In a German TV-interview in the beginning of the sixties, she made a similar remark. At the same time, however, she showed that she would like to be treated by others as a philosopher. Respect for Hannah Arendt need not go so far as agreement in her opinion that she is not a philosopher. No one who confines his reading to *The Human Condition* would think of calling the author a *political* philosopher. It does not diminish my respect for the work of Hannah Arendt to say that I cannot consider *The Human Condition* a substantial contribution to political philosophy. Politically, Hannah Arendt may well defend freedom and plurality in political space, but these political achievements were confined to the Athenian aristocracy in ancient Greece. They paved the way for that elite's achievement of "worldly immortality," when their deeds were thought worthy of inclusion in the chronicle of history. When the chips are down, we are probably standing on the wrong side.

> The distinction between man and animal runs right through the human species: only the best (*aristoi*), who constantly prove themselves the best (*aristeuein*, a verb for which there is no equivalent in any other language) and who "prefer immortal fame to mortal things," are really human; the others, content with whatever pleasure nature will yield to them, live and die like animals. (19)

Most of us belong to the second category and will miss this political crown of immortal fame. The only fortune granted us will be that we will now and then be witnesses of revolutionary transformations—events unpredictable with regard to their origin and outcome, unpredictable events in which people act as political actors in Hannah Arendt's sense. One may have Hungary 1957 in mind, to mention one of Hannah Arendt's own examples, the GDR 1989, South Africa 1990, USSR 1991.

Margaret Canovan has illuminatingly described the conflict between the elitarian and radical-democratic ideas in Hannah Arendt's work.[2] She emphasizes a crucial distinction that is sadly lacking in Arendt's thought.

> The distinction is that between what one may call *normal* politics and *extraordinary* politics, and it is unfortunate that the same concern for rare events that gave her unparalleled insight into extraordinary politics should have let her overlook normal politics. The theory of politics as the unexpected, unpredicted actions of a few free men is an excellent account of what happens in extraordinary political situations.

Thereupon Magaret Canovan very kindly remarks: "While her inconsistencies need to be recognised, however, it is important not to make too much of these defects in her thought."[3] I absolutely agree but nevertheless still object to politics as an institutionalized space for rivalry (*aristeuein*)!

The importance of *The Human Condition* lies, in my opinion, not in Hannah Arendt's contribution to political philosophy, but in the philosophy of science and technology that reflects on the implications of science and technology for our culture. For a reason that for me is still a riddle, most books dedicated to her neglect this aspect.

The Human Condition begins and ends with the philosophy of culture characteristic of Hannah Arendt. Her first sentence comes straight to the point: "In 1957, an earth-born object made by man was launched into the universe, where for some weeks it circled the earth according to the same laws of gravitation that swing and keep in motion the celestial bodies—the sun, the moon and the stars." What then happened she calls "an event, second in importance to no other!" As a philosopher, she reflects on the cultural implications. She looks at this event from the perspective of man escaping from his imprisonment "to" the earth. Her discussions of "labour," "work," and "action," so central in *The Human Condition*, are to be interpreted as panels for her criticism of modernity in the finale of her book.[4]

Alienation from the World

Hannah Arendt defends the uncommon claim that people in modernity are alienated from the world. This is uncommon, because most of us find secularization, utilitarianism, consumerism, hedonism, materialism, and so on characteristic of this time, and in these words the concentration on life's daily worries and pleasures is reflected in a number of different ways. What has Hannah Arendt in mind when she speaks of modernity's alienation from the world?

World in her idiom is a typically human construction and is contrasted with the cyclical natural processes of rising, shining, and decaying. When Hannah Arendt speaks about this world, it is not the physical world she refers to. Her concept of world separates human beings from and protects them against nature. World is the human artifice of objects and institutions that guarantee them a permanent and durable home. Man is naturally artificial. According to Hannah Arendt, not the natural but the artificial is specifically human. Civilization gives man the opportunity to transcend the animal species and consists precisely in building a world: a world of ploughed fields, roads, and hedges instead of a wild landscape, a world of buildings instead of the open air, a world of language and culture, of communities and traditions, a world of art, law, religion, and all the rest of the man-made things that outlive the men who made them and form the inheritance of the human race.[5] This creation is more permanent than the individual and represents a certain stability for him. Each new generation inherits this specifically human and relatively stable context and adds her part to the cultural web that she hands down to the next generation.

Marx is convinced that modern man has been alienated from *himself*. Hannah Arendt denies this. The problem of our time is that man has been alienated from the modern world in the sense that this world no longer represents for him the touchstone of reality. Man has become detached from his artificial inheritance: his experiences have been reduced to subjective, private ones. The world is no longer his common heritage, but a whole being in permanent transformation. The world does not offer man a stable dwelling-place, with the result that he experiences himself as worldless. In this way his security has been undermined and his life deprived of meaning. He cannot share his life-experiences with others. What is permanent in modern life is the hunting for the new that relieves the guard of the old.

The present article discusses the role of science and technology in this process of world alienation. It is Hannah Arendt's view that the common world—the world as artificial and durable—has been overpowered by the cyclical life process to which man is bound by necessity and which threatens to absorb all manifestations of human-ity.[6]

The History of Physics

At the close of *The Human Condition*, Hannah Arendt discusses three charac-teristic events on the threshold of modernity: (1) the discovery of America and the ensuing exploration of the whole earth, (2) the Reformation which started the twofold process of individual expropriation and the accumulation of social wealth, and (3) the invention of the telescope and the development of a new science that considers the nature of the earth from the viewpoint of the universe. These three events put moder-nity in a characteristic light.[7] Following Hannah Arendt herself I shall not discuss the first two phenomena but focus upon the new science.

As regards the history of physics, Hannah Arendt explains her own view, which differs from Werner Heisenberg's. Certainly, she has studied his *Das Naturbild der heu-tigen Physik*, but she ignores his "phases" in the history of physics. The difference be-tween Hannah Arendt and Werner Heisenberg amounts to the fact that she takes the characterizations of modern physics from the twentieth century, as described by Werner Heisenberg, as though they were the profile of physics as such, classical and modern as well. To clarify the difference I shall first discuss Heisenberg's view.

After the foundation of physics in the seventeenth century, the view of nature dominant in the Middle Ages changed. Nature was considered foremost as the crea-tion of God. Kepler also shared this viewpoint. In that the scientist was still more en-grossed in the particular occurrence of nature, he was led by the judgment that the details of nature could be isolated from their connections, mathematically described, and explained. The tendency became still more clearly apparent that nature was to be considered independently not only of God but of man as well. This amounted to the ideal of describing and explaining nature objectively. Newton's decisive step was his insight that laws of mechanics applying to falling stones also determine the movements of the moon. In this way the meaning of the word *nature* as an object of research changed: it became a container concept relevant for all those domains of experience which man could penetrate by means of science and technique, irrespective of nature being accessible to direct experience. In the nineteenth century nature was considered a happening in space and time governed by physical laws, in principle regardless of nature's description by man and his intervention. In our century profound changes have taken place in that the behavior of the smallest materials, particles, cannot be discussed without taking into account the way in which they are observed. The observation of the particle is as such a disturbance. The result has been that the laws of nature mathematically formulated in quantum physics do not apply to particles as such, but to our knowledge of the particles. The question whether these particles exist in space and time can no longer be posed in this form. The representation of objective reality has in a remarkable way evaporated. Modern science always presupposes a human being. He is not only spectator but also co-actor in the theater of life, as Bohr put it. Heisenberg thinks that the modern physical situation may be simplified in a not too flagrantly unjust way if one contends that for the first time in history man encounters only himself. The elementary particles cannot be considered the last objective reality, they withdraw themselves from an objective arrest in space and time, with the consequence that we can only objectify our knowledge of these particles. In other words, in science the object of research is no longer nature but nature exposed to our examination. In that sense man encounters only himself.

The great difference between Werner Heisenberg's representation of physics and Hannah Arendt's is that she shortens the historical perspective in such a way that the features distinguishing physics from the seventeenth century and physics from our own century disappear. She says: "The modern astrophysical world-view, which began with Galileo, and its challenge to the adequacy of the senses to reveal reality, have left us a universe of whose qualities we know no more than the way they affect our measuring instruments" (261). Here I emphasize the astrophysical view beginning with Galileo and the challenging of the senses. More pronounced is her deviant statement that modern relativism has its parentage not in Einstein but in Galileo and Newton (264).

Hannah Arendt does not see Galileo as a physicist looking at objective reality but as somebody with an astrophysical view of reality. Galileo used the telescope in such a way that the secrets of the universe were delivered to human cognition "with the certainty of sense perception." That is, he puts within the grasp of an earthbound creature and its body-bound senses what had seemed forever beyond his reach (260). According to Hannah Arendt, the telescope did not mean an extension of human senses. In a certain sense the telescope deprived the senses of their reliability, because it brought their insufficiency to light. Hannah Arendt attaches Galileo's name to, on the one hand, the ancient fear that our senses, our very organs for the reception of reality, might betray us and, on the other, still more important, to the realization of the Archimedean wish for a point outside the earth from which to unhinge the earth (262). For whatever we do today in physics, we always handle nature from a point in the universe outside the earth. Without actually standing where Archimedes wished to stand, still bound to the earth through the human condition we have found a way of acting on the earth as though we had it at our disposal from outside, from the Archimedean point. This is Hannah Arendt's frame of reference for speaking of the alienation of the earth by science. In the scientific experiment man realized his newly won freedom from the shackles of earthbound experience. Instead of observing natural phenomena as they were given to him, he placed nature under the conditions of his own mind, that is, under conditions won from a universal, astrophysical viewpoint, a cosmic standpoint outside nature itself (265).

As compared with Werner Heisenberg, Hannah Arendt makes her own classification of physics. She finds a difference between science which looks upon nature from a universal standpoint and thus acquires complete mastery over her, on the one hand, and a truly "universal" science, on the other, which imports cosmic processes into nature even at the obvious risk of destroying her and, with her, man's mastery over her (260).

The Rise of Cartesian Doubt

While the new science, the science of the Archimedean point, needed centuries and generations to develop its full potentialities, taking roughly two hundred years before it even began to change the world and to establish new conditions for the life of man, it took no more than a few decades for the human mind to draw certain conclusions from Galileo's discoveries and the methods and assumptions by which they had been accomplished. The human mind changed in a matter of years or decades as radically as the human world in a matter of centuries (271). That is why Hannah Arendt focuses her attention on the father of modern philosophy, although she emphasizes that the author of the decisive event of the modern age is Galileo rather than Descartes, because not ideas (Descartes) but events (Galileo) change the world. In this context she reminds us that the heliocentric system as an idea had a long history to show for itself, but never changed the world or the human mind (273).

In modern philosophy and thought, doubt occupies the same central position as did the Greek *thaumazein* in ancient philosophy, the wonder at everything that is as it is. Cartesian doubt was originally the response to the new reality where not reason but a man-made instrument, the telescope, actually changed the physical worldview. It was not contemplation, observation, and speculation which led to new knowledge, but the active stepping in of *homo faber*. If neither truth nor reality is given, they must be conquered by the activity of making and fabricating. There is nothing left to be taken on faith. *De omnibus dubitandum*, says Descartes. Everything must be doubted. This doubt doubts that such a thing as truth exists at all.[8]

Descartes's philosophy was haunted by two nightmares. In the first place reality— of the world as well as of human life—was doubted. It might be a dream. In the second place his senses and reason could not be trusted. An evil spirit, a *Dieu trompeur*, might willfully and spitefully betray man. The question of certainty was at stake. The loss of it ended in a new, entirely unprecedented zeal for truthfulness. In science nature was trapped with experiments and instruments so that she would be forced to yield her secrets. Instead of the theory revealed through a contemplative glance it was a question of the hypothesis practically tested. The success of this test became truth.

The Cartesian resolution of doubt, or its salvation from the two nightmares, was similar in method and content to science. It may be interpreted as a turning away from truth to truthfulness and from reality to reliability. If there is no truth, man can be truthful; if there is no reliable certainty, man can be reliable. If there was salvation, it had to lie in man himself; if there was a resolution of doubt, it had to come from doubting itself. If everything has become doubtful, then doubting at least is certain and real. The famous *Cogito ergo sum* was in essence a *Dubito ergo sum*. It was the introspection of consciousness that yielded certainty.

Of great relevance was the fact that the Cartesian method of securing certainty against universal doubt corresponded to the most obvious conclusion to be drawn from the new physical science: though one cannot know truth as something given and disclosed, man can at least know what he himself makes (282). This became the generally accepted attitude of the modern age, and this conviction propelled one generation after another for more than three hundred years into an ever-quickening pace of discovery and development (283). What men now have in common is not the world but the structure of their minds. In a certain sense Descartes placed the Archimedean point in the human mind (284).

Scientists formulate their hypotheses to arrange their experiments and use their experiments to verify their hypotheses. They deal with a hypothetical nature, a "physical reality." In other words, the world of the experiment seems always capable of becoming a man-made reality, and this unfortunately puts man back once more into the prison of his own mind, into the limitations of patterns he himself created (288).

The conclusion Hannah Arendt draws is not that truth and knowledge are no longer important, but that they could be won only by "action" and not by contemplation. The reasons for trusting "doing" and mistrusting "contemplation" and "observation" became even more cogent after the results of the first active inquiries. Scientific truth need not be eternal, it need not even be comprehensible or adequate to human reason! (290). Thinking—traditionally thought was conceived as an important way to the contemplation of truth—became less relevant than doing, whereas contemplation, in the original sense of beholding the truth, was altogether eliminated.

For the modern philosopher it meant that he turned away from the world of phenomena and from the world of eternal truth and withdrew into himself. What he discovered in this region of the inner self was the constant movement of sensual perceptions and the activity of the mind. Most of modern philosophy since then became theory of cognition and psychology.

The Process

The most important spiritual consequence of modernity's discoveries had been the reversal of the hierarchical order between the *vita contemplativa* and the *vita activa*. The fundamental experience behind this reversal was the thirst for human knowledge that could be assuaged only after man had put his trust in the ingenuity of his hands (290). This thirst was inspired not by a pragmatic desire to improve human conditions on earth, but exclusively by an altogether non-practical search for useless knowledge! Within the *vita activa* the activities of making and fabricating became of utmost importance, with the result that *homo faber* occupied the leading position on the cultural stage. This did not last long, however, because according to Hannah Arendt *homo faber* had to leave his place to *animal laborans*. This shift of position was a social replication of what happened in science. Hannah Arendt explains this phenomenon as follows.

An important transformation took place in science when the question of what things are was converted into the question of how things work. Instead of the ontological question (what is this?) the problem of the functionality of the process (how does it work?) became central. Ernst Cassirer has described the same climate change in epistemology by his *Substanzbegriff und Funktionsbegriff*.[9] Nature as a being became nature as a process. Whereas it was in the nature of a being to appear and thus to disclose itself, it is in the nature of a process to remain invisible, to be something whose existence could be inferred only from the presence of certain phenomena. This shift from substance to function undermines the activities of *homo faber* for whom the production process was a mere means to an end. The new emphasis makes the process more important than the product. The implications of this reversal of means and ends remained latent so long as the mechanistic worldview, the worldview of *homo faber par excellence*, was predominant. In the mechanistic climate the instrumentalization of the world could come to full maturity—a world full of instruments and artificial objects subjected to the useful ends of *homo faber*.

The esteem of *homo faber* was quickly followed up by the elevation of laboring to the highest position in the hierarchical order of the *vita activa* (306). *Homo faber* made place for *animal laborans*. The modern shift from the "what" to the "how," from the thing itself to its fabrication process, undermined the position of *homo faber*. It deprived man as maker and builder of those fixed and permanent standards and measurements that had guided him. The fame of *homo faber*—building a world and fabricating worldly things—was blown away in the wind of modernity, characterized by alienation from the world and by introspection. In this context Hannah Arendt speaks about the role played by the development of commercial society that first introduced the principle of interchangeability, then the relativization and finally devaluation of all values (307). For the mentality of modern man, determined by the development of modern science and philosophy, it was at least as decisive that man began to consider himself part and parcel of the two superhuman processes of nature (Darwin) and history (Hegel/Marx). The ultimate failure of *homo faber* is indicated by the rapidity with which the principle of utility—the very quintessence of his worldview—was superseded by the principle of "the greatest happiness of the greatest number."

Secularization has not been inspired by love for the world, after the evaporation of the transcendental world of religion and metaphysics. Modern man was thrown back upon himself and not upon his world. When he lost the world to come, he did not gain this world. Modern man was thrown into the closed inwardness of introspection, where the highest he could experience were the empty processes of reckoning of the mind, its play with itself. The only contents left were appetites and desires, the senseless urges of his body. The only thing that could now be potentially immortal was life itself, that is the possibly everlasting life process of the species Mankind.

When Hannah Arendt compares the modern world to the past, there is no reason for merriment. The victory of *animal laborans* is almost complete. The trouble with modern theories of behaviorism is not that they are wrong but that they could become true as the best conceptualization of certain trends in modern society. It is quite conceivable that the modern age may end in the deadliest, most sterile passivity. These theories do not articulate human action, or work, but exclusively labor as part of the life process! And the danger is that people will think about themselves in these terms, abandoning their individuality and acquiescing in a tranquilized functional type of behavior.

But there are further dangers. In her essay "The Conquest of Space and the Stature of Man" she also writes about the danger that man sees himself from the perspective of the Archimedean point:

> Seen from sufficient distance, the cars in which we travel and which we know we built ourselves will look as though they were, as Heisenberg once put it, "as inescapable a part of ourselves as the snail's shell is to its occupant." All our pride will disappear into some kind of mutation of the human race. The whole technology seen from this point no longer appears as the result of a conscious human effort to extend man's material power, but rather as a large-scale biological process. Under these circumstances speech and everyday language would indeed be no longer a meaningful utterance that transcends behaviour even if it only expresses it, and it would much better be replaced by the extreme and in itself meaningless formalism of mathematical signs. The conquest of space and the science that made it possible come perilously close to this point.[10]

In *The Human Condition* these are not Hannah Arendt's last words. She ends with a palliative: Needless to say, this does not mean that modern man has lost his capacities or is on the point of losing them. Men persist in making, fabricating, and building, although these faculties are more and more restricted to the abilities of the artist. Similarly the capacity for action is still with us.

The cultural implications of science and technology have been drawn into a strong perspective. Hannah Arendt clearly shows that tasting science and technology changes man and his culture. I should like to conclude with a number of considerations.

Evaluation

Hannah Arendt's Interpretations

1. *Distinctions in Science and Technology.* As was mentioned above, Hannah Arendt makes her own classification in the history of natural science as compared to Werner Heisenberg's. I am doubtful as to how felicitous Hannah Arendt's implicit critique of Heisenberg's phases in classical and modern physics is. Unconvincingly she sweeps all characteristics of classical and modern physics together, including Einstein's theory of modern relativity, and makes Galileo's telescope the crucial factor.

Hannah Arendt could say that, in the light of modern physics, classical physics can get, and has got, a new meaning, because in modern physics appears what has been enclosed in classical physics. In that spirit one could defend the view that also in classical physics man encounters himself, in the sense that the representation of classical physics is not to be gathered in everyday reality, but represents a specific human design. What man used to call the "objective" reality was his own construction. Nevertheless, due to this construction, man never becomes a co-actor in the theater of life. In other words, nature does not depend upon man in the sense Bohr originally meant. Simply neglecting the stages sketched by Heisenberg does not yet open up the possibility of assigning historical meaning to the telescope in the way Hannah Arendt does. From the historical perspective the telescope is overloaded.

Another, in my view interesting, distinction of Werner Heisenberg's which Hannah Arendt regrettably passes over refers to the history of technique itself. In the eighteenth and nineteenth centuries the development of technique rested upon the use of mechanical procedures. This form of technique was, in the first place, a continuation of the old craftsmanship. According to Heisenberg, it was not the steam engine which altered this development, but electronics, in the second half of the nineteenth century. In electronics there was no longer any relationship with the old craftsmanship. It was a matter of exploiting natural forces unknown to man in intercourse with nature up to that moment. Also chemical technique entered into the old craftsmanship of dyeing, of pharmacy, etc., but around the turn of the century chemical technology took new paths. And it was from scratch that in atomic technology natural forces were developed for us, to which people had no access on the basis of everyday experience.

For practical and theoretical evaluation of the various techniques, these two distinctions may be helpful. In my opinion Hannah Arendt gives them too short shrift. The so-called conquest of space makes a deep impression on her, as if this step meant an incision *par excellence* in human history. On the one hand, she evaluates it as a progressive change in the history of technology and, on the other, as a continuation of astrophysical thinking nestling on the Archimedean point. But the success of the voyage to the moon has nothing to do with modern physics. Nor was this voyage the result of an unmanageable process, but chiefly a *homo faber*-like activity, in which a clear target was set.

2. *Interpretations in Philosophy.* A footnote must be made to Hannah Arendt's interpretation of Descartes. Just as she projects the insights of modern physics onto Galileo, she attributes to Descartes himself the modern experiences of Descartes's great-grandchildren. It is historically rather improbable that Descartes formulated the philosophical answer to the challenges of classical physics.

In the first place, Descartes's contemporaries and friends, like Mersenne and Arnauld, drew his attention to parallel thoughts about doubting in the work of Augustine. The church father, too, departs from sceptical doubt in order to come home via the *cogito*.[11] In any case, the copyrights on doubt of the truthfulness of the senses do not belong to the *homo faber* of the telescope, nor to the father of modern philosophy. This does not mean that I like to put Augustine and Descartes on the same footing. The refutation of skeptical doubt is to be sure the same, but Augustine does not base his whole philosophy and theology on this insight. The words are the same, but the point is nevertheless different. Indeed Descartes broke with his predecessors, insofar as he showed the ambition to begin with the beginning without putting his trust in the authority of the philosophy that went before. Moreover Descartes did not like to confuse what was clear and distinct with conjectures or at best probabilities. His philosophy was new not so much in content as in method. "But as regards all the opinions which up to this time I had embraced, I thought that I could not do better than endeavour once for all to sweep them completely away, so that they might later on be replaced either by others which were better or by the same when I had made them conform to a rational scheme."[12] His ideal had been inspired not by classical physics but by the mathematics in his time. According to Dijksterhuis, a historian of science, Descartes always operated in a mathematical way, in his metaphysics as well as in his physical works.[13] Hannah Arendt herself gives the key to the understanding of her position in a footnote where she remarks that Kierkegaard gives us the deepest penetration of Cartesian doubt (275). In other words: Just as she sees Einstein rising behind the back of Galileo, she projects Kierkegaard upon Descartes. But neither Descartes nor Kierkegaard mentioned the Archimedean point laid by them in man. This point of reference for reality is not their creation.

Eccentricity

Hannah Arendt interprets the rise of science as the discovery of the Archimedean point from which the earth is to be unhinged. This metaphor is not a felicitous one; it suggests a rather static state of affairs. It gives the impression that after the discovery of that point, first scientists, then philosophers, and finally all people look upon the human world from this *removing* center. In accordance with this, Hannah Arendt defends her thesis that science necessarily implies alienation from the earth. Physics is called astrophysics. Her assumption is that man belongs to the earth, and in scientific projects she seems to detect man's rebellion against the earth and the body-bound senses, in short this rebellion against his own condition. She certainly articulates a very important and interesting point of view with regard to the science of astrophysics and its alienating dangers.

I prefer to defend another view on these matters, in the wake of the German philosopher Helmuth Plessner.[14] He contends that alienation belongs to the human condition; by definition it is his nature. This could still be in line with Hannah Arendt. But his theory continues as follows: Man has the faculty of transcending his context and even himself. In transcending he is enabled to consider his own position. Man is, so to say, a double (*Doppelgänger*). He is himself, and at the same time he is a spectator of himself. He is his body, and at the same time he has a body. Being his body represents his "central" position; that he can look at his own body represents his "eccentric" position. In this way man moves permanently between these two positions, centric and eccentric. In the centric position man keeps contact with the world from his center. The boundaries of this center are not fixed forever. To begin with they coincide with the skin of the body, but there exists the possibility that this body acquires some *extensions*. The blind man explores his environment with his white stick and the astronomer the celestial space with his telescope. Driving in a car, the driver feels the bumps of the road through the body-work. These instruments, the coachwork included, are to be considered as bodily extensions and may become part of one's body. One may feel at home with them if one coincides with these extensions.

In the eccentric position man transcends himself and this enables him to look at himself and his behavior. It does not end with looking; he can also act out of this eccentric position. In my view the centric and eccentric positions form the extremes of the scale of possibilities. Often man is in the middle of these extremes: fully occupied in his business (centric position) and at the same time conscious of what he is doing (eccentric position). It is this eccentric position that represents a form of alienation: he who is eccentric is not in his own center, he is not at home. Being outside himself he *considers* himself. In this theory, alienation is not a phenomenon of modernity, but a constitutional feature of human beings. Philosophy as an expression of reflection is a personification of alienation.

Eccentricity represents a position from which man can look at himself, but also one from which man can explore or even act. Anthropologically, so-called astrophysical knowledge is possible only through the faculty to move to an eccentric position. The creation of the world in the way Hannah Arendt means is also possible due to this eccentric faculty. Man transcends his natural context and becomes an outsider of his natural environment, so that the world he creates is an alienation of nature. Anthropologically, alienation is an ingredient of the human condition, because man is and always remains an outsider to himself. This condition is permanently and structurally accompanied by unrest and restlessness. This unrest and restlessness urge man to transcend the world he has created and made into his dwelling place. He then prepares for leaving the world he had nestled in and will thus meet a new alienation.

I make abundant use of typically Plessnerian terms, such as eccentricity, centricity, transcendence, alienation, etc., but I have to admit that Plessner's anthropology is different. In Plessner's work, eccentricity is characteristic of the difference between man and animal. Animals do not know this faculty of transcendence. According to Plessner this structural moment being the border line between man and animals represents, technically, a transcendental motive. (1) My own understanding of this intriguing concept is a little different. I am less interested in the characteristic difference between man and animals. For myself the concept is not useless if biological research produces evidence in support of a gradual form of eccentricity among monkeys. Moreover the distinction between centric and eccentric in man is not absolute; both positions form the extremes of a scale one can also be in the middle of. This is not an orthodox Plessnerian view. (2) I am interested in the history of humanly eccentric *positions* because I am convinced that modern society is more eccentric than traditional societies. The Jewish belief in the transcendental God, for instance, is an expression of a more eccentric position than the magic-animistic faith of a primitive tribe. To avoid misunderstandings, I do not look at the development of eccentric positions in terms of its own history as an ascending line. In short: if eccentricity is a transcendental category, as Plessner supposes, then there is no possibility of being in the middle, between the extremes (cf. [1]) and eccentricity has no history, because it is structural (cf.[2]).

The Archimedean Point

It is a matter of a circular process of creation and recreation due to human eccentricity. Man creates a world in which he wants to dwell, but the internal human split is not healed. He frees himself in order to bind himself. His liberation always ends up in captivity. Each human project goes this way. Hannah Arendt considers the Archimedean point as a center moving away from us human beings, and she thinks that the dangers lie in this removal. I propose to interpret the Archimedean point not as a center but as an eccentric position. Because the double of centricity and eccentricity is a human specification, it is not in Descartes the Archimedean point is found for the first time. Whoever has developed a certain sensitivity discovers the Archimedean point, in Augustine, in Archimedes, in short in the whole of the philosophical tradition where it is a matter of reflection. And not only there!

This introduction of eccentricity as structural alienation aims at putting Hannah Arendt's view on astrophysics into perspective and also relativizing it, without intending to throw all eccentric positions on one heap. On the contrary, I consider Hannah Arendt's plea as a strong argument for the view that, on the threshold of modernity, eccentricity grows in intensity with the rise of science. In my view Hannah Arendt overreacts to the manifest demonstration of eccentricity in science—overreacts because of her conviction that the human senses as such are deprived of their value by the eccentricity of astrophysics.

Only if one renounces one's own sensual experiences may a mental space arise in which another counterintuitive—i.e., scientific—perspective can flourish. If one imagines this precondition, it is ridiculous to call the scientific approach objective in the sense of unquestionably right. But without doubt, in scientism, positivism, etc., it is a matter of the colonizing of everyday experience by science. If one accepts that there is no argument in favor of talking about objective reality in classical physics, then one can take for granted that people can approach reality in different ways, for instance in a scientific and in an everyday way. Everybody makes use of these possibilities, not at the same time but alternately. What is right in a scientific context need not be right in another context, and vice versa. The everyday approach will undoubtedly be influenced by science, with the result that the everyday approach of three centuries ago is no longer the same, but the strong influence of science on the value of the senses which Hannah Arendt suggests seems not quite right. More strongly, the distrust of the senses at the expense of everyday experience—developed within the context of classical physics—is not only systematically but also historically groundless. Accordingly, the metaphor of the telescope is not very felicitous. It did not deprive the human senses of their function. The telescope shows what you cannot see with your naked eye; in other words, it extends sight. Historically it was also appreciated in that way. The same can be said of the microscope. It is a means for my eye, even if a new world is opened up. The fascination of telescope or microscope may have the consequence that one overlooks what is near at hand. But if one looks at what is near at hand by means of a telescope one does not use one's eyes very well. Whoever walks in the forest with an astrophysical look loses his sense of proportion. That is more than an imaginable danger of the modern technological culture, where the technological imperative is more important than the categorical.

Technological and Scientific Culture

It is not my intention to restrict the cultural meaning of science and technology by saying that they only represent a new approach, parallel to the everyday approach. The relevance of philosophical reflection on the cultural meaning of science and technology has not only to do with approaches. Science and technology show a tendency to conjure up a culture of their own, contrary to culture up till now. It is a culture built upon "astrophysical" insights. These insights may not take man in his bodily and earthly experience into full account. What conflicts will arise as a result of these "astrophysical" insights, one sees in practice where what is scientifically and technologically possible—being the answer to problems that do not yet exist—is brought into practice. These technical possibilities are the logic of the new world. Not every medical-technological possibility is helpful for human beings.

Science and technology, creating new artifacts, have decisively modeled the face of the world, and this new face of the world gives shape to human thinking, feeling, and action. In this circle of modeling and shaping, "astrophysical" knowledge circulates, i.e., knowledge not directly related to concrete human experience and to an earthly measure. These artifacts mediate this specific scientific and technological culture. Their influence is probably much stronger than the radiation of the scientific discourse held in terms of astrophysical points of view. Nevertheless one has to be on one's guard against equating the technological universe and the everyday lifeworld—conceptually and practically.

This viewpoint of eccentricity offers an interesting possibility for thinking about astrophysical knowledge and its products. They are not *as such* inimical to men. Many artifacts show the possibility of being incorporated in the earthbound human body. After intensive training a car can become an artificial extension of the human body with which we can move without any difficulty. That was not the case when we prepared ourselves for the driving test. Sitting behind the wheel for the first time we felt uneasy, troubled, awkward, nervous, etc. The car did not do what we wanted it to do. After a learning process the car becomes incorporated and the steel-plated car has become our second skin. When we learned to drive a car we did it from an eccentric position. In this learning process we moved from an eccentric position to a more centric position, the car becoming a part of ourselves. It goes without saying that a philosophical reflection on technological artifacts is not allowed to confine its attention to the viewpoint of incorporation. A psychology of the drivers and of the walker can make clear how much the human world changes under the influence of the so-called incorporated car. A driver's relationship to his fellow drivers under way becomes more aggressive, his perception of the landscape becomes more global, etc. In other words the application of eccentricity to the phenomenon of the car does not imply a plea for considering the car, from an evolutionary point of view, as desirable a part of ourselves "as the snail's shell to its occupant." This provocation is interesting for the purpose of clarifying the dangers that threaten us. The dangers only increase when they are not spoken about.

I have attempted in the above to create some distance to Hannah Arendt's view on science and technology, trying in this way to create space for discussing the dangers. Probably the most dangerous problem does not lie in the necessary alienation resulting from science and technology, but rather in the alliance of science and technology with the economic process. It is in this alliance that the phenomenon Hannah Arendt describes so adequately in terms of the scientific and technological *process* gets its rigidity. Speaking about the double structure, or eccentricity, the new alienation acquires utopian traits. But that is a matter to be discussed separately.

Notes and References

1. Hannah Arendt, *The Human Condition* (Chicago: University of Chicago Press, 1958), un-prefixed page references are to this work.

2. Margaret Canovan, "The Contradictions of Hannah Arendt's Political Thought" *Political Theory* 6, no. 1 (1978): 21.

3. Ibid., p. 23.

4. By *labour* she means that activity human beings have in common with animals and is dictated by biological necessity. Through *work* an artificial world is evoked. These artifacts are durable creations (i.e., appliances, artistic creations, buildings, monuments) and are the permanent *coulisses* for human life. Hannah Arendt considers *action* the highest human activity. In the sphere of action people show who they are. In her book *The Human Condition* this articulation of the *vita activa* is illustrated with ancient Greek material. With the same categories she analyzes modern society. These concepts registering remarkable shifts in the *vita activa* necessarily demonstrate the decline and alienation of modern society in comparison with the Greek polis.

5. Cf. Margaret Canovan, *The Political Thought of Hannah Arendt* (London: J. M. Dent & Sons, 1974), p. 81.

6. It is interesting to reflect on the influence of Hannah Arendt's political experiences in the thirties on her ideas about modern worldlessness in modernity. Cf. my article "Hannah Arendt als bewuste Paria," *Wijsgerig Perspectief* 32 (1991/92), nr. 1.

7. It is not her aim to offer a classic treatise of causes and effects in connection with our situation. It is rather our situation that sheds light on the past. These three characteristic events have been crystallized into modernity.

8. "[T]he traditional concept of truth, whether based on sense perception or on reason or on belief in divine revelation, had rested on the twofold assumption that what truly is will appear of its own accord and that human capabilities are adequate to receive it. That truth reveals itself was the common creed of pagan and Hebrew antiquity, and of Christian and secular philosophy" (p. 276).

9. Ernst Cassirer, *Substanzbegriff und Funktionsbegriff: Untersuchungen über die Grundlagen der Erkenntniskritik* (Berlin: Verlag Bruno Casirer, 1923).

10. *Between Past and Future* (Berlin; New York: The Viking Press, 1974), p. 279.

11. *De Civitate Dei*, XI 26. Cf. Etienne Gilson, *Etudes sur le role de la pensée médiévale dans la formation du système cartésien* (Paris: J. Vrin, 1930), ch. II, in which Gilson also mentions another locus where Augustine refutes radical skepsis.

12. Descartes, *Discours de la Methode*, in Adam and Tannery (eds.), *Oeuvres de Descartes* (Paris: Leopold Cerf, 1898), VI, pp. 13–14.

13. E. J. Dijksterhuis, *De mechanisering van het wereldbeeld* (Amsterdam: J. M. Meulenhof, 1975), p. 447. Copleston also remarks that it is not clear what role Descartes attributes to experiment and experience. Frederick Copleston, *A History of Philosophy* (New York: Doubleday, 1960), vol. IV, p. 81.

14. Helmuth Plessner, *Die Stufen des Organischen und der Mensch* (Berlin: De Gruyter, 1928), and *Zwischen Philosophie und Gesellschaft* (Frankfurt: Suhrkamp, 1979). Very important is Petran Kockelkoren's thesis on Plessner, *De natuur van de goede verstaander*, WMW-publikatie 11, Universiteit Twente, 1992.

PART VII

The Human and the Non-Human

U NLIKE PHILOSOPHY OF science, which is closely identified with science itself, philosophy of technology has long struggled against the growing technologization of modern life. Perhaps something important about the nature of technology is overlooked from this negative standpoint. This in any case is the thesis of these concluding chapters which highlight French contributions to the field.

Paul Dumouchel's chapter "Gilbert Simondon's Plea for a Philosophy of Technology" introduces a thinker who is widely viewed as the major French philosopher of technology. Simondon focuses on "the mode of existence of technical objects" and the processes of their evolution and development. His work offers an original account of the nature of modern technology, and of the alienation that results from its growing automatism. Simondon believes that humanistic nostalgia and hostility to technology result from the decline of traditional technical culture in which the human hand and eye coordinated the use of tools. A new technical culture must reconcile us once again with the mechanical world we have created. That culture, Dumouchel concludes, will have a surprising relevance to ecology.

Bruno Latour's "A Door Must Be Either Open or Shut: A Little Philosophy of Techniques" develops a whole theory of the relation of the human to the non-human in the course of explaining a simple cartoon. The clever hero of the piece uses technology to reconcile the conflicting wishes of his colleagues and pets. Latour shows how to analyze this example in terms of the place of technology at the meeting point of "programs" and "anti-programs" reflecting different social demands. His account shows the inextricable interconnection of humans and things in the organization of social life. The attempt to separate them, to elaborate a social theory without techniques, and a technology without humans, is doomed to failure.

15

Gilbert Simondon's Plea for a Philosophy of Technology

Paul Dumouchel

Is THERE AN essence of technical objects? Do they form a natural kind? Does our classification of certain things as technical objects carve nature at the joints or is it purely nominal, a simple convention? One classical paradigm of a technical object is the tool, and classificatory conventions are often considered as convenient tools. If this is revealing about the nature of some classifications, can it tell us anything about the nature of a tool? Following David Lewis (1969) perhaps all conventions can be viewed as tools of some sort, tools of coordination. Some tools then are conventions. Are all technical objects conventions? Presumably, a hammer is not, at least not in the way money is a convention. I mean that not just its value is conventional, but the fact that it is money rather than a mere piece of paper or gold is also conventional, not natural. No physical property ever determines that an object is or is not money, or that it is a conventional object: a convention does. Should we believe this is also true of an antibiotic or a spring mattress?[1] There are two questions here. Does the fact that something is a watch or an electric motor correspond to a set of physical properties of that object? Is there a necessary and sufficient physical characterization of an object which allows us to determine that it is a technical object? A positive answer to the first entails nothing about the second question. It may be that being a technical object is like being an adaptation, the result of a particular history. The classification of a trait as an eye or a wing rests on its physical characteristics, but the fact that this eye, or this wing, is an adaptation depends on its history, on how it evolved, not on its physical properties.[2] Likewise the classification of a substance as a neuroconnector depends, among other things, on its physical properties; but the fact that this molecule is a biological or a technical object would depend on its history, on the way it was produced. Should we conclude then that our distinction between natural and technical objects cuts no ice, that it can tell us nothing about the nature of an object, about what it is, its physical properties which are causally active in the world? Is technology a pseudo-concept and are the problems which, we say, it causes not mere illusions but badly misrepresented when we assign responsibility for them to something which has no reality of its own?

Or is it the case that our distinction reveals a kind of dualism of properties and that technical objects can be successfully characterized by some non-physical, perhaps supervenient, properties. Properties which, like adaptation or self-organization, are objective determinations of existent realities.

I find it revealing that what is commonly called the philosophy of technology pays so little attention to such typically philosophical questions. Most of what goes by that name centers its reflection on the social and political consequences of technology or the assessment of technological risk and the impact of new technologies. It is true that many labor under an impression of impending doom and extreme urgency, but it may be that, as philosophers, we are putting the cart before the horse. Do we know what we are talking about when we say 'technology' or 'technical objects'? It was his conviction that we do not, which, more than thirty years ago, led Gilbert Simondon to write his classic book in philosophy of technology: *Du mode d'existence des objets techniques*.[3] I think that what he had to say is still relevant in the 1990s. In the remainder of this chapter I will present some of Simondon's central ideas on technology. I will not be able to do justice to all of his insights in such a short space, but I hope, as he did, to succeed in stimulating some philosophical curiosity about the nature and mode of existence of technical objects.

Simondon believed philosophy had to fulfill a duty of understanding toward technical objects similar to what the enlightenment accomplished toward slaves and primitive people. There is no naive anthropomorphism here, just the belief that ignorance begets oppression and understanding, freedom. This belief, or at least its second half, is often challenged today and our recent history is taken as proof that it must be false. That the growth of scientific knowledge and modern technology has led to unprecedented political and ecological disasters is offered as irrefutable evidence of the failure of Kant's ideal of the enlightenment. This is a point of view brilliantly defended by George Steiner (1971) among others. It could be that Steiner is the victim here of a philosophical prejudice well entrenched in our tradition. Since Galileo and Hobbes, at least, it is common to think we understand what we make better than anything else. The art of the craftsman who assembles a complex object from its elements is taken as a model of analysis and explanation, as an epistemological ideal. We understand more thoroughly what we make because in this case we are, so to speak, the authors of the object of our representation. Technology, then, is applied science, and technical objects wholly transparent artifacts whose sole reality is in the design and intention of those who conceive them. Thus it is knowledge, science, and understanding, rather than the lack thereof, which have failed us and betrayed the hope the eighteenth century naively placed in them.

I, for one, suspect we understand what we make no better than we understand what we do, and since action has received intense scrutiny revealing an unsuspected depth of philosophical problems, we can doubt that technology is knowledge incarnate.

Use and Technology

Common understanding, and some philosophical reflections, have it that the proper way to approach technology is through the concept of use, or of instrumental rationality, the means/end relationship. The classical paradigms of instrumental rationality are economic behavior and tools, instruments made to serve one or a variety of purposes. This has led to the idea that technology has essentially the structure of labor, or work, man's relation to nature, and it has raised interminable debates surrounding the issue of the neutrality of technology.[4]

According to Simondon, the analysis of one or a few simple examples suggests that this identification of technology with instrumental rationality, through the concept of use, is, to a large extent, misleading. Take a watch or a clock: the use or function of such objects is that of keeping time. A watch which needs to be rewound from time to time, with a spring and a pendulum, is essentially a spring motor, a contraption technically related to a crossbow. An electric watch or clock hides behind its dial an electric motor, an object technically related to an electric buzzer, and to many doorbells. As for the digital electronic watch, it functions on technological principles radically different from those of the two preceding items.[5] Social demand, or use, jumbles together in misleading categories completely heterogeneous technical objects "for no fixed structure corresponds to a particular use" (1958, 19). A watch is not a well-defined technical object, though it may be a representative specimen of a species of commodities. The language of use, which is to a large extent the language of exchange and commerce, does not map onto the technical realities of the objects exchanged or used. It constitutes an elementary black-box theory of technology which defines a telephone or a battery by its input-output functions. (Until some deadly stuff starts oozing out, the innards are ignored.) We are generally oblivious to the variety of structures which similarity of use conceals. The way a technical object is used tells us little about whether it belongs to a natural family, whether it is part of a class in which all members share common physical characteristics. The subsumption of technical objects under the category of use ignores their technical characteristics, their technicality. If its internal structure is what makes a technical object what it is, the concept of use bypasses the essential.

This is related to a current doctrine in philosophy of mind and cognitive science, known as functionalism, and to the issue of realism in epistemology. According to functionalism, highly different material or physical structures can support the same functional organization which defines one machine. Conversely, a plurality of machines, or systems, can be defined over a unique material structure. In fact, one system can be defined for every use we may dream of, each machine corresponding to a specific functional organization. Is it, then, that Simondon lacked the concept of functional organization and hence failed to define a machine at the right level of abstraction? Should we be Marxists, then, and believe that the market discovers (produces?) the true universality of the concept of a machine, which eludes the technician imprisoned in particularity?

Simondon was well aware of the concept of functional organization but his point concerns the reality of technical objects. A machine defined by its functional organization, Simondon would argue, is strictly an abstract logico-mathematical object rather than a technical one. Current wisdom has it that both brains and computers are universal Turing machines, but we should not forget that a universal Turing machine is neither a biological system nor an electronic device. Consequently it is unlikely that this concept of a machine will be able to capture the difference, if there is one, between a biological and a technical object. This is not surprising: the idea of functional organization has been essentially used, in philosophy of mind, as a means of erasing, as much as possible, the difference between certain technical and biological systems, i.e., computers and brains. This is a legitimate research strategy, but if our goal is to inquire into the nature of technical objects, it seems rational to avoid, at first, conceptual analogies which mask their specific difference.

The idea of functional organization is a theoretical construct and there are doubts, in many quarters, about the existence of such entities. A soft drink vending machine can be viewed as a cognitive system which recognizes genuine American or Panamanian coins, as an automated vending machine, as a cumbersome doorstop; or, during large family gatherings, it can be laid on its side to serve as a bench. The physical states of a machine underdetermine its functional states. According to Daniel Dennett (1989), there is no fact of the matter, outside of the interpreter's or Mother Nature's intentional stance, which determines what such an object is, what its functional states are. Should we believe then that automated vending machines exist only from a certain point of view, that they have no independent reality? Yet underdetermination is not absence of determination. There is some fact of the matter that a telephone book is not an internal combustion engine. It is an objective fact which no stance or point of view can erase. That is why Dennett's rider, concerning Mother Nature, is so fundamental. Evolution produces material structures which perhaps underdetermine but nonetheless determine that something is a leg or a horn. Through the study of these structures, we can retrace the evolutionary history that led to their present form and discover whether they are adaptations or not. Thus we come to recognize real patterns in the world which result from a causal process. These patterns constitute objective realities about which there is some fact of the matter.[6] Simondon claims that the same is true of technical objects. They have a reality which is independent of the user's stance and which can be discovered by studying their history and evolution.

The Mode of Existence of Technical Objects

Consider an internal combustion engine, that of a car or a motorcycle: the most primitive examples of this technical device, according to Simondon, can easily be recognized through an analysis of their structure and technical organization, without any information concerning the date of their production. Early representatives of a technical lineage are abstract objects. Abstract, in the sense that the assemblage of material parts giving a physical existence to the mental representation of a function[7] and of the interaction of various processes lacks integration. The processes are concatenated, or sequentially related, but they are rarely interdependent; the realization of one process does not rest on the virtualities and potentialities of another. The first tokens of a technical object still bear the mark of its mental origin, of the analytical separation of realities which, like color and shape, exist *partes extra partes* only in the mind. As it is refined, this artificiality disappears from the object's successive representatives and is replaced by an interdependency of the functions and processes which reveals, as much as it reflects, the structures of the physical world. Simondon calls "concretization" this evolution characteristic of technical objects. It is the historical process through which an abstract schema enters into the concrete world of material things and physical processes. According to him, concretization is not the mere meeting of an ideal form and a completely amorphous matter, but a process of discovery which progressively exploits the virtualities of the physical world in the context of an abstract schema. That is why the level of concretization defines for Simondon the degree of technicality of a technical object and can be used as a measure of technical progress. Thus the concept is both normative and descriptive: it measures a progress and describes an evolution. Three innovations in the internal combustion engine may help to illustrate this process.

The first is the passage from the water-cooled to the air-cooled engine. The water-cooled engine uses an independent cooling system which is only indirectly linked to the motor, for example, through a belt which activates the water pump so that the water circulates faster when the engine works harder. If ever this belt breaks, as may well happen, the engine will be damaged since the conditions which ensure its proper functioning are no longer satisfied. The important point is not so much that the engine will be damaged, something that can happen to any technical object, but the fact that the various conditions which allow the proper functioning of the engine are distributed over nearly entirely independent subsystems. In an air-cooled engine, the cooling rests on the radiant heat of the exposed parts and on the convection process thus created. This means that the very functioning of an air-cooled engine creates the cooling process which allows its proper functioning (1958, 22).[8] The air-cooled engine takes advantage of a necessary aspect or dimension of the processes taking place within the engine and assigns it an analytically independent function. In a way the evolution of technical objects resembles biological evolution, tinkering with potentialities present in the material and processes involved.

The second technical innovation is the invention of the diesel engine. In a regular gasoline engine the explosive mixture of air and fuel is ignited by a spark provided by an independent electrical system. In a diesel engine the mixture spontaneously detonates when certain conditions of heat and pressure, provided by the functioning of the engine, are satisfied. This last statement is rigorously true of any internal combustion engine. However, the diesel engine exploits a phenomenon: the fact that a volatile mixture will spontaneously ignite under certain conditions, the occurrence of which is prevented as much as possible in a regular gasoline engine. Its occurrence in such engines is an undesirable and destructive effect known as pinging. The diesel engine, argues Simondon, constitutes progress because it realizes a more complete integration of the physical processes at work during the cycle. It is less abstract. It takes advantage of an ever-present potentiality whose actualization must be prevented in a regular gasoline engine, though the conditions which determine its occurrence are, necessarily, repeatedly approximated by the functioning of the engine. Thus the diesel engine constitutes a more stable solution to a problem of internal coordination, and represents a better dynamic equilibrium of the system of natural processes mobilized by the technical object. The mental representation has taken a particular form, determined by the potentialities of the material world.

Our third innovation concerns the shape of the cylinder head of an air-cooled engine. In order to ensure both a better diffusion of heat, by increasing the heat-exchange surface, and more efficient convection in the air surrounding the engine, the cylinder head has received a characteristic shape. It resembles a column on which circular blades mounted on thicker rings and slightly separated from each other have been slipped. At first the blades were as if they had been properly slipped over the cylinder. Its thickness was calculated to resist distortion under the internal pressure and the blades were added on to this ideally predetermined volumetric unit. In more recent motors, the blades play a dual role. On the one hand they serve their original function of facilitating the cooling process. On the other hand, they serve a mechanical role. They act as structural ridges which allow thinner walls to withstand greater pressure. Simultaneously the thinner walls permit a better diffusion of the heat so that the two functions of the blades are inseparable. If one were to cut off the blades of such an engine, not only would part of the cooling system be destroyed, but the walls would no longer be able to stand the internal pressure. Two abstract functions, a thermodynamic process of heat exchange and the mechanical rigidity of a surface, which are analytically separated (and preferably so) in the mental representation of the engine, are now concretely integrated in a unique material structure.

Through concretization the abstract mental representation of a technical object is progressively embodied in a definite structure of materials and physical processes. According to Simondon, it is best thought of as a teleological process, but its end point should not be seen as the conscious goal pursued by scientists and engineers in the refinement of technical objects. Simondon's claim is not an empirical hypothesis about the psychology of invention. Like spontaneous orders according to Hayek (1973) concretization is the result of human actions but not necessarily of human design. In fact, the model of teleology at work here seems imported from physics.[9] The actualization of potentialities in a technical system partially resembles the reduction of potential energy in a physical system. It is a more stable state, one toward which the system will tend. Though we as technicians may fail to notice certain potentialities, or, as users, may prefer the greater acceleration of the regular gasoline engine, a technical essence has a propensity to evolve toward a state of equilibrium of the physical processes it involves.

When the original abstract schema is sufficently rich, concretization can lead to the appearance of a technical individual. An air-cooled engine cannot function without, as a consequence, simultaneously cooling. A diesel engine cannot function if the proper conditions for the ignition of the fuel are not fulfilled, but the functioning of the engine fulfills those very conditions. These two characteristics are of paramount importance in understanding technical individuals. Greater subsystem interdependence gives rise to a more autonomous object. Technical individuals depend less on external subsystems because they integrate in a network of circular causality more of the processes that determine their proper functioning. They thus acquire a form of circular, or self-referential, organization reminiscent of biological systems. The result, a technical individual, is a physical system whose functioning determines, to a large extent, but never entirely, as we will see later on, the proper conditions of its functioning.

To say that something is a technical object, therefore, is to say that it results from a particular history characterized by a definite type of evolution or development.

The Evolution of Technical Objects

This notion of technical progress renders the evolution of technical objects independent of social demand and the pressure it exerts upon the distribution and modification of such objects. Concretization responds to a different set of imperatives than the one which determines the use of a technical object. The internal requirements of reciprocal adaptation, as indicated above, are sometimes at odds with the demands and desires of users. This suggests a difficulty linked to the conflation of a normative and a descriptive notion in the idea of concretization. I will argue later on that concretization is a good local measure of technical progress, but I think that its relation to technical evolution is not quite as simple as Simondon appears to indicate. The fact that social demand can hinder technical progress is a clear sign that concretization cannot be equated with technical evolution. At least, if we want to retain that term to designate the appearance of new technical objects and change over time in the distribution of such objects, concretization is only one of the forces which shape technical evolution. As such it is more like natural selection or genetic drift in biological evolution, not evolution itself, but an evolutionary force the effect of which can be reduced or increased by the presence of other forces. Thus, Simondon's research points to the need for wider inquiry, to identify and track these other forces and to understand how they act conjointly in the evolution of technology.

Concretization is teleological and that constitutes another reason why we should be wary of identifying it with technical evolution. According to Simondon every technical lineage has an absolute origin which it is in principle always possible to determine (1958, 40–49). Conceptually, all motors are related, those which use the same type of energy, like fossil fuel, even more so. Yet your car's motor is (most probably) not an improved steam engine, nor a descendant of such a device. The transition from external to internal combustion necessitates a radically new design, a different abstract schema. The internal combustion engine mobilizes different physical processes and establishes new types of interactions among them. As such it opens up a new set of possibilities on which concretization capitalizes. For Simondon, the beginning of a technical lineage is marked by "a synthetic act of invention which is constitutive of a *technical essence*. The technical essence is recognizable through the fact that it remains stable throughout the evolutionary lineage, and not only stable, but also productive of structures and functions by internal development and progressive saturation" (1958, 43). Concretization is the historical process which tends toward the integration of all the potentialities included in the essence of a lineage and emergent from the physical processes which constitute the object. It follows that the transformations any technical lineage can sustain are limited and converge upon one or a few fixed points, states of dynamic equilibrium, beyond which the technical object can hardly be improved. Concretization resembles ontogenesis more than evolution, and it is interesting that Simondon uses both terms interchangeably. It consists more in the actualization of a predefined set of possibilities, which reaches an ultimate state of saturation and dynamic equilibrium, than in progressive adaptation to an external environment.[10]

It follows that concretization provides only a local measure of technical progress. Simondon considered that technical individuals were more advanced than simple tools because they carried the process of concretization further. From that point of view, concretization seems like a global measure of technical development. The difficulty is that concretization is always determined in relation to the the abstract scheme which constitutes the essence of a technical lineage and that comparisons across lineages necessitate an evaluation of the technicality of the different schemas. Unfortunately, concretization provides no clues to the evaluation of schemas and even suggests that the technicality of abstract schemas is an inconsistent idea. In consequence, it appears that the domain of application of concretization, as a measure of technical development, is limited to ordering successive representatives within a technical lineage. What should we conclude from all of this?

If we want an objective measure of technical progress then it should be independent of social demand, of changing fads and fashions, and of the whims of users. In fact, if no fixed structure corresponds to a definite end, the notion of technical progress should be independent of the use to which we destine technical objects.[11] These are the precise characteristics of concretization, and Simondon has provided, I believe, a valuable local measure of technical progress. Though it needs to be refined and extended, it constitutes a step in the right direction.

Concretization escapes the sociological pitfall of identifying technology with its various uses. It also avoids the intellectualist mistake of equating technical progress with complexification of conceptual design or functional organization. A technical object is not a mere mental representation. It is a concrete material thing, the seat of various physical processes and potentialities. These material conditions make possible the evolution of technical objects, and it is the material existence of such objects which makes them relevant to us as things with which we can physically interact. The original design of a technical object, which systematically relates various physical processes, is in a way similar to concretization. But the attempt to embody an abstract mental representation as a concrete material object leads to continued invention and discovery in which materials offer unforeseen resistance or possibilities and the interaction of known processes produce previously unknown effects. Concretization implies a form of technical judgment and intuition which cannot be equated with theoretical knowledge. This is precisely what we should expect of a good notion of technical progress. It should take into account not only the conceptual design of an object but also its technicality: the interaction of the various physical processes it involves, the evolutionary possibilities they contain, and the limitations they impose. It should also point toward a specific cognitive activity which, like aesthetic judgment or moral reasoning, is neither entirely separated from nor wholly identical to pure reason and theory.

Thus Simondon has provided a valuable local measure of technical progress. He has identified one of the major forces in the evolution of technology and charted out some of the main differences between technical and biological evolution. Not a mean achievement.

Hypertelic Tools and Technical Individuals

Major objections can still be raised against Simondon's conception of technology and technical development. One is that if concretization is well tuned to the analysis of motors and machines, it seems less appropriate for the understanding of simple tools, like a shovel, or of complex technical ensembles, like a building site, a shipyard, or a factory. Another is that technical objects are made to be used, they supply a social demand, and it is through their use and application that they transform the social and natural world. One cannot therefore simply relegate use to the minor role of a perturbation which distracts technology from its normal development toward greater concretization. The relation between technology and social use is too important to be so easily dismissed. The two objections are closely related, as we will soon see, for those technical realities which seem less amenable to an analysis in terms of concretization are also those in which use plays, apparently, a more important defining role.

It was no mere accident that Simondon centered his analysis of technical objects on technical individuals. He thought they revealed a central aspect of technology, also present, but somewhat obscured, in tools and technical systems. Furthermore, he believed that the changing and difficult relationship which now exists between human agents and technology stems to a large extent from the accrued presence of technical individuals in our societies. Thus his attitude was influenced partly by pedagogical demands, partly by what he saw as one of the fundamental problems of his subject matter.

Take a simple tool like an adze. It is, in fact, a relatively complex technical object, not simply a homogeneous piece of metal cast in a certain form. Its chemical composition varies from point to point. It has been forged, which means that the molecular chains have received a definite orientation at different places. If it is to be a good tool, its blade must be sharp and remain so even when working hard woods. This implies that the cutting edge should be composed of steel, but the steel should extend only over a well-defined area, otherwise the blade will become brittle and break. The part which rises immediately behind the cutting edge and becomes progressively thicker should be made of a material which retains sufficient elasticity to fulfill its function as a wedge and a lever in the wood. Thus, it is "as if the tool in its totality was made of a plurality of functionally different areas welded to each other" (1958, 72). A good tool is one in which the problems of interaction between these functionally distinct parts have been harmoniously resolved. In a machine the coordination of the physical processes involved is continuous and contemporary to the functioning of the machine, while in a tool the integration of the functionally distinct parts is achieved, once and for all, when the tool is manufactured. Yet, in both cases, the process is essentially the same: concretization. The actualization of certain potentialities leads to a state of equilibrium between the various requisites. In both cases, it entails a reduction of the margin of indetermination between the processes involved. Not every chemical composition or orientation of the molecular chains will result in a good adze. The refinement of the tool explores an open set of possibilities until it converges upon a more stable and satisfactory solution, one which implies a more determinate relationship between the functionally distinct areas of the adze, just as the passage to the diesel engine entails greater precision in determining the moment when the fuel is injected in the cylinder. This loss of indetermination through the actualization of potentialities, whether in a tool or a machine, leads to a more autonomous object. An object which provides to a greater extent the conditions of its proper functioning or use.

Simple tools are also subjected to a different direction of development. They can acquire a specific characteristic which augments their adaptation to one of their various uses, or to a limited subset of them. For example, a hammer can gain a convex surface in order to push the nail below the level of the wood. In consequence of this improved local adaptation the hammer becomes misadapted to numerous tasks for which it previously constituted a perfectly adequate tool. For evident reasons, Simondon names this development hypertelic adaptation [*hypertélie ou adaptation hypertélique*]. The object becomes highly or exceedingly finalized. The hypertelic tool or element has a limited domain of use or application.[12] As such it will be more easily damaged, more often useless. The range of situations in which it is misadapted is wider. It is therefore not surprising that hypertelic objects are usually found as components of larger technical ensembles in which they have a specific use or function. Within these ensembles the conditions of their proper use are duly satisfied. Technical ensembles provide a regulated environment in which tools are not simply exploited but cared for and employed only in appropriate circumstances. Until recently, human agents were the central piece of most technical ensembles. They were the ones who accomplished this indispensable regulation. That occupation gave rise to a classical technical culture, in which use of the proper instrument for each operation and care of one's tools were paramount values of the craftsman. According to Simondon we have been able to play this role because we are autonomous individuals.

In technical individuals the external regulation provided by human agents becomes, to a large extent, self-regulation. This is why such objects are named technical individuals. It is also why they can replace us as bearers of tools. Their self-regulation can be extended to the regulation of hypertelic instruments within wider technical systems. According to Simondon, the rise of technical individuals destroyed our ancient technical culture: by replacing us as bearers of tools, the conditions which gave meaning to our technical culture's values were destroyed.

Hypertelic adaptation reveals that the technical object exists at the interface of two different worlds: the world of its own internal requirements and the world in which it is used. Thus technical objects undergo two types of adaptation which often pull them in opposite directions.[13] According to Simondon, the craftsman acted as a mediator between the demands expressed by consumers and the technical requisites of the objects trusted to his care. This mediation was expressed in a technical culture which exerted a profound influence on social and political organization. It shaped the medieval city and structured a complex hierarchy of crafts and trades. Technical individuals also adapt to the world in which they are used. This adaptation, without being hypertelic, ensures their continued existence. It is a relation somewhat like that of a living organism to its environment. According to Simondon, the criterion for the existence of a true technical individual is the presence of an associated niche [*milieu associé*]. This niche is a portion of the natural world transformed to provide an adequate environment for the technical object. It is not only an adaptation to the world but also an adaptation of the world for, and sometimes causally by, the technical individual. The best example that comes to mind is a hydroelectric dam. Considered as a whole it consists in a complete individual which provides not only the internal conditions of its own proper functioning but its appropriate environment. The associated niche is that part of the natural world which is transformed in order for the technical individual to exist as a functional reality. It is a technico-geographical environment, neither entirely natural nor entirely artificial, an area or domain of the natural world which has been permanently altered in view of providing a satisfactory context for the functioning of a given technical individual. Likewise, the electric grid which covers our countries constitutes the associated niche of thousands of technical individuals. It not only provides them with a necessary source of energy; many are absolutely dependent on the 50hz or 60hz cycle of the alternating current.

Technical individuals pose novel problems of coordination in larger ensembles to prevent destructive interactions between their associated niches. These problems are at the heart of our current predicament concerning modern technology. They are difficult problems and we now address them, argues Simondon, without an adequate technical culture. Technical individuals have rendered our classical technical culture obsolete. In their self-regulation and the transformation of nature, they also embody a new type of adaptation to the world in which they are used. Technical individuals substitute, to a large extent, direct communication between various natural processes for human mediation embedded in a social organization. That is why it may be that modern technology exerts a much more superficial influence on the structures of our societies than is commonly believed. Simondon suggests that our difficult relationship to technology is not unrelated to that situation.

Rarely today is one born in a dynasty of tanner, coppersmith, musician, or farmer, and if one is, one can walk away without forfeiting one's best social assets. Among the many reasons which explain the progressive disappearance of such occupational lines, the establishment of market economy has certainly been central, but the appearance of new technology has been just as important. The much lamented lack of skills and absence of vocation characteristic of the modern worker is to a large extent inseparable from his unprecedented freedom of choice. We need not do as our forefathers did, because our tools ask so little of us. A rapidly learned trade, though it may fulfill an essential social function, is not a worthwhile heirloom, for the excellent reason that there is no way of preventing it from being learned by anyone. Modern technology has not only transformed our countryside and the face of our cities, it has also destroyed particular social organizations which attached, often from his birth to his death, a man to his tools. We may complain that this new freedom has left us with little sense of direction, but we should remember that this novel situation rather spells, as Simondon suggests, a lesser, not a greater, technological determination.

It is, Simondon thought, because we are freed from our tools that we are so estranged from the technology which we make. In a society in which we no longer occupy the place of technical individuals, we tend to meet technical objects only in the commodities which house them. We lose sight of their technicality and reduce them to their use. Their reality escapes us. It is completely misunderstood. We see in them mere tools whose true meaning is in our intention and purpose.[14] Alternatively we construe technical objects as unnatural things, as if their technicality were not a historical quality but an intrinsic property of the processes they mobilize. In conclusion, I would like to indicate some of the consequences of Simondon's view for our understanding of the relationship between technology and nature and the relationship of his conception to two major philosophical intuitions concerning the essence of technology and its meaning, Hegel's and Heidegger's.

Nature's Virtualities and Their Actualization

A few years ago a Montreal newspaper reported the results of a province-wide survey in which people were asked to rank what they considered to be the most damaging natural catastrophes which could hit the province of Quebec in the near future. The responses overwhelmingly identified a general electrical blackout as the foremost natural disaster which threatened the province, followed in second place by a powerful earthquake.[15] For a Marxist or a Hegelian this answer reveals the extent to which agents are alienated in the modern world. To consider an electrical blackout as a natural catastrophe is to place the results of our own activity outside ourselves as a foreign nature which dominates us. It is a failure to recognize ourselves in the products of our own activity. One should not blame nature, but the utility company. For the Heideggerian, such a response can be seen as a symptom of the era of "total mobilization," which unwittingly reveals the essence of technology.[16] In its naïveté and irrationality it bears witness to the fact that we now see nature only as a "fund," as a reserve of resources to be exploited by us, and that only such resources exist for us as nature. The response betrays that the essence of technology is not something which is technical, but a mode of presentation of being. Simondon's view indicates, I believe, that, in this case, common understanding may be closer to an analytically correct intelligence of technology than part of our philosophical tradition suggests.

Lightning and electric eels can astonish us, and at times terrify us, but very little can be learned from them concerning the laws of nature[17] which govern the behavior of one of the four major forces of the universe. It would certainly be awkward to consider this fundamental reality as something artificial, yet it is clear that most of the laws which constitute the core of the science of electricity, like Joule's laws, were inferred from studying artificially produced phenomena. They measure currents which do not exist "naturally." The phenomena these laws regulate have no being outside of the technically constructed situations in which we produce them, just as some transuranium elements may have no other existence than their fleeting appearance in a cyclotron.[18] In that sense an electrical blackout is a "natural" catastrophe. There is no electricity which is "natural" in opposition to the technically produced energy which heats our homes in the dead of the Canadian winter.

To some extent Simondon agrees with Heidegger: modern technology is a "mode of presentation of being," but it is not a mere way in which what is appears and it does not define an epoch in the history of being, "the end of metaphysics." Technology gives existence to virtualities contained in reality and "brings out into the open" new events and processes. It is not a pure creation but a way of progressively discovering the world's capacities. What makes modern science so different from its ancient Greek predecessor is that through technology it creates the phenomena which it saves.[19] Thus we should not be surprised that when the associated niche of technical individuals mobilizes already existing environments which give rise to unprecedented processes, these are often destructive of the established order or equilibrium of what is. This is not due to the class origin of our science, or to the mercantile intentions of those who make technical objects, nor is it the consequence of the fact that modern technology is a total mobilization of all that is for the illusory satisfaction of our goals; the disturbing aspect of modern technology is a clear consequence of its creative dimension. It would be extremely surprising if the irruption of a novel process which brings into existence a hidden virtuality of reality naturally harmonized with the already existing equilibrium of processes in which it appears. It is not because of the uses to which we put it that modern technology radically transforms the world, but because technology gives existence to phenomena which were not there before and because technical individuals provide the conditions of the processes which constitute them. Thus there is no alternative technology which contains different values which respect nature. What technology teaches us is that there is no "nature" in the sense of a set of events and processes which are essentially different from those which are produced artificially. According to Simondon, there is no technology which can respect what is, for technology is essentially the coming into existence of the virtual.

It does not follow that we should abandon all the ecological or social questions technology puts to us, but it may be that we should consider them in a different manner. Simondon invites us to see that there is no unnaturalness in technology and that what we call "nature" is in part the product of our own activity. This, to some extent, is a Hegelian view and it is not uninteresting to compare the two approaches. Hegel thought that through its labor humankind would come to recognize nature as its own product and no longer posit the results of its own activity as an alien reality which dominates its makers. We should meditate Hegel's intuition. Despite its grand metaphysical dress, it refers to a strangely familiar reality. Our recent tools have made us responsible for many processes which were, before, safely abandoned to the self-regulation of nature. Forestry, birth-control, and determining the sex of children are excellent cases at hand. Until very recently population regulation, reforestation, and the determination of the sex of our children existed as purely natural processes. Though through magic we sometimes dreamed of controlling them, they were not under our care and responsibility. They belonged to the domain of things which, according to the Greeks, appear of themselves, *phusis*, nature. They are now objects of our policies, subject to our laws and open questions for our ethics. In this sense Hegel and Simondon agree, there is no longer a nature independent of us and which is not in some way the result of our own activity. Through the product of our labor, technology, we have put an end to our alienation from the world and have become responsible for what we previously considered as a foreign reality to which we were subjected.[20]

It is nonetheless an illusion to consider that this is a particular consequence of modern technology. Pre-classical Greece was a country covered by forests and which received abundant rainfalls. Then trees were cut to make boats and to clear new areas for the culture of wheat, necessary for the fast-growing population. Unfortunately the ancient Greeks did not know that once it reached a certain threshold, this deforestation would radically alter the climate of their country. Unwittingly, they created the dry, semi-desert peninsula which we know today, and the sudden drop in the yield of the wheat fields forced them to colonize Sicily, the island which came to be known as the granary of the Mediterranean. What, it seems, is determining is not the passage from ax to chain saw but the size of our intervention in an environment. Time and the number of the participants will easily compensate for the little power of the instruments involved. What modern science and technology have given us is not so much an unprecedented capacity to transform "nature" as the means and the knowledge to assume responsibility for our actions. We cannot be forced to take responsibility for our actions, but we must face their consequences and are compelled to deal with them.

In many traditional societies tanners formed a caste of untouchables who were exiled to the outskirts of cities. Their trade involved constant intimacy with death and carcasses and their workshop gave off an unbearable stench. Nonetheless, all wore shoes, rode on saddles, carried bags. This irresponsible attitude is called today the "not-in-my-backyard syndrome." Free-riding, we should remember, has little to do with the essence of technology or the defense of the environment. It is a bad craftsman who blames his tools.

Simondon believed philosophy could help promote a new technical culture to replace our current regimen of fear, hatred, and irrational hopes. These attitudes, he thought, followed from lack of reflection. We still have a long way to go before we understand technical objects. He sketched the map of an immense philosophical continent.[21]

Notes

1. This is the direction in which some sociologists of science seem to be going, for example Bruno Latour (1989) at least as I understand him. I think that much confusion would be cleared if we took pains to distinguish the different questions which are involved here. Even if no physical property of an object ever determines that it is a technical object it does not follow that no physical property determines that it is a steam engine even if steam engines are technical objects.

2. It is true that adaptation is usually thought to be supervenient on physical properties but this means precisely that adaptations are not physical properties. This dualism of properties is also apparent in the distinction between evolutionary homologies and evolutionary analogies, respectively: similar structures which share a unique evolutionary history and similar structures which do not share an evolutionary history. Similarity of structure is not a sure sign of a shared phylogenetic history; hence evolutionary homologies can be different adaptations.

3. First published in 1958, it was reprinted in 1969 and again in 1989 by Aubier-Montaigne, Paris.

4. The dialogue between Marcuse and Habermas is a good example of this type of approach.

5. We may add to this list of different technical structures fulfilling the same use or function: solar dials, sand clocks, candles and water clocks.

6. I do not pretend that Dennett's doctrine concerning real patterns claims exactly this, but I am tempted to believe that it should.

7. The function of an internal combustion engine is understood here as its functional organization, as the thermodynamic exchange, the work it produces and the order in which the different parts intervene in the complete cycle of the engine, rather than as the different uses to which it can be put like producing electricity, or pumping air in order to drive a sledge hammer, or moving a car or an airplane.

8. It is true that in some air cooled engines mounted on mobile devices, like cars or motorcycles, the movement of the air consequent to the movement of the device plays at times a major role in the cooling of the engine, but in other devices like an electric generator or a lawn mower it plays virtually no role at all.

9. There is good textual evidence, especially in Simondon (1964 and 1989), which supports this interpretation.

10. Not all modern conceptions of biological evolution are at odds with Simondon's view of technical evolution. For example, Elredge and Gould's (1977) theory of punctuated equilibria strikes me as conceptually related to Simondon's analysis, inasmuch as it rests on the idea of species selection and adopts a saltationist stance. The main difference is that species do not have "absolute origins" and that the phylogenetic tree is related from its base to its furthest branches.

11. This does not mean that we should do without the concept of a function. The function of a motor may be to produce mechanical energy and the function of a hammer, percussion, but these should not be confused with the various uses to which we can put these objects.

12. Hypertelic adaptation can affect any technical reality whatsoever, a motor or a machine as well as a tool or a technical element, but a hypertelic machine will not be a true technical individual.

13. Clearly, this is not always the case, especially when the two tendencies act at different levels. The hypertelic adaptation of a subcomponent may in fact facilitate the integration of two processes within one technical individual.

14. It does not really matter, I think, whether we see in technology a neutral reality whose value is determined by its use or if we view it as embodying a definite (political) purpose indifferent to our best intentions. In both cases its technicality is obscured and its reality reduced to (subjective or objective) intention.

15. In order to understand why a general electrical blackout can be considered as "most damaging" one should consider first the climate of the province of Quebec which is extremely cold in winter and the fact that such blackouts have a history of being very long, often more than twenty-four hours.

16. About the notion of total mobilization see Heidegger (1977) and E. Jünger (1930).

17. The concept of law of nature has recently been criticized by various authors (Cartwright 1983, 1990; Van Fraassen, 1989) and, as we will see, Simondon's view also suggests a reevaluation of this notion.

18. On the idea of the technological construction of the phenomena studied by modern science, see Bachelard (1951).

19. On this topic see Bachelard (1951), Arendt (1958), and Prigogine and Stengers (1979).

20. From this point of view, much critique of technology is but the expression of fear and lack of direction in front of new-gained freedom and responsibilities.

21. This research was made possible thanks to the generosity of the Fonds FCAR of the Province of Québec, Canada, the Social Sciences and Humanities Research Council of Canada, and the hospitality of the CREA, Paris, France. I also want to thank Mary Baker, Andrew Feenberg, and Lucas Sosoé for learned advice and useful suggestions.

References

Arendt, H. (1958). *The Human Condition*, University of Chicago Press, Chicago.

Bachelard, G. (1951). *L'activité rationaliste de la physique contemporaine*, Union Générale d'éditions, Paris.

Cartwright, N. (1983). *How the Laws of Physics Lie*, Cambridge University Press, Cambridge.

Cartwright, N. (1990). *Nature's Capacity and Its Measurement*, Cambridge University Press, Cambridge.

Dennett, D. C., (1989). *The Intentional Stance*, MIT Press, Cambridge MA.

Elredge, N., and S. J. Gould (1977). "Punctuated Equilibria: The Tempo and Mode of Evolution Reconsidered," *Paleobiology*, pp. 115–51.

Hayek, F. A. (1973). *Law, Legislation and Liberty*, University of Chicago Press, Chicago.

Heidegger, M. (1971). *The Question Concerning Technology*, Harper & Row, New York.

Jünger, E. (1930). *La mobilisation totale*, Gallimard, Paris, 1990.

Latour, B. (1989). *Science in Action*, Harvard University Press, Cambridge, MA.

Lewis, D. (1969). *Convention: A Philosophical Study*, Harvard University Press, Cambridge MA.

Prigogine, I., and I. Stengers (1979). *From Order to Chaos*, Bantam Books, New York.

Simondon G. (1958). *Du Mode d'existence des objets techniques*, Aubier-Montaigne, Paris.

Simondon G. (1964). *L'individu et sa genèse physico-biologique*, Presses Universitaires de France, Paris.

Simondon G. (1989). *L'individuation psychique et collective*, Aubier-Montaigne, Paris.

Steiner, G. (1971). *In Bluebeard's Castle: Some Notes towards the Redefinition of Culture*.

Van Fraassen, B. (1989). *Laws and Symmetry*, Clarendon Press, Oxford.

16

A Door Must Be Either Open or Shut[1]

A Little Philosophy of Techniques

Bruno Latour
Translated by Charis Cussins

THERE IS NO better way to think about the essence of a technique than through a simple example—at least, that is our bias as empirical philosophers. And so as not to intimidate the reader with cutting-edge technology, let's consider the invention of a door by that master of invention, Gaston Lagaffe, Franquin's cartoon hero. In one cartoon strip, everything is said: the essence of a technique is the mediation of the relations between people on the one hand and things and animals on the other.

"Miaow!" There's a cat mewing in the office of the magazine *Spirou*. What's a cat doing in a Belgian office? We won't linger over this question. Whatever the reason, the cat is mewing and demanding that Prunelle, Gaston's boss in the office, open the door.

"I've become a doorman for cats," exclaims Prunelle, indignant at having been mechanized, instrumentalized, coopted by a door, by a cat, and by Gaston. Just like existing dedicated door closers—human or mechanical[2]—Prunelle has become a human open-and-shut-door mechanism. His stiff, furious posture (second frame) indicates sufficiently that he is imitating a machine, acting like a robot. Crisis point is quickly reached, triggered by the cat's incessant mewing. It wants the door to be open all the time so that it can come and go freely. Prunelle ought to know this about cats. His ignorance annoys Gaston. "Don't you know that a cat can't bear shut doors?!. . . . and that it needs to feel free?!" Gaston—speaking for the rights of cats—and the cat—worthily represented by Gaston and also capable of expressing itself by its soul-wrenching miaows—thus want the open-and-shut-door mechanism to be on guard at all times to respect the rights of animals. Walls and doors are meaningless to felines, and while they want to take advantage of the comforts of the hearth, they don't want to be prisoners to it. Perfect parasites, they want to take everything and give nothing. Domesticated but wild; such is the cat.

But this ignores human rights, and in particular those of Prunelle, to protect himself from drafts. Drafts! How many disputes they provoke in buses, trains, offices! People will kill for an open or shut window. And yet it seems that drafts kill only the French and the Belgians. The British, for example, don't claim that drafts badly threaten their health. But the Belgian Prunelle has to weigh in the balance the psychology of this "poor dear" of a cat, and his health. The first demands that doors be open, the second that they be shut. If felines remain wild, the journalists are civilized, and stay warm. Prunelle's manner shows that he hasn't the slightest hesitation on this point. "These doors will be closed," he proclaims, using the future tense of commands, and he underlines his categorical imperative with the growl "rogntudj," a comic strip symbol of authority, and by the "schlam" of the doors that he slams shut in fury. There is nothing more to be said. Cats and subordinates must obey.

That's to reckon without the technical gesture, however: the ruse, the detour, the *daedalion*, the *metis* (stratagems); to reckon without the patched-together bricolage, in which, since the dawn of time, the ingenuity of Daedelus, Vulcan, or Gaston Lagaffe has been found.[3] "Each time Prunelle pesters me, I've found a trick to stop him being the strongest," mutters Gaston, now equipped with a saw and toolbox. Its the millenary wisdom of the engineer that our new Archimedes is invoking. Take a boss with authority and power. Oppose him with an engineer who has nothing but a handful of tricks. Who alters the power relations? The engineer, of course, as we've known since Plutarch. "King Hieron," wrote Plutarch, "*sunnoesas les tecnes ten dunamin* (amazed by the power of techniques), ordered war machines from Archimedes for the defense of Syracuse" after having seen him pull a trireme full of men all alone, with the help of the little trick of his composite pulleys.[4] Archimedes redefined power: an old man, a rope, and some pulleys become stronger than a trireme team and a sovereign who talks loud and strong.

Gaston, more modestly, only redefines doors and invents (or reinvents) the cat-flap: "a practical little opening at the bottom of a door that lets cats come and go," as it says in the dictionary. Gaston's cat-flap is a vertically opening door set in a horizontally opening door. The hinges replace our friend Prunelle, who no longer has to act as cat doorman. The mechanized human gave way to an automatic mechanism. The translation through which the human groom became a machine groom was done by the mediation of the hinges. Instead of the continuous presence of Prunelle, Gaston only had to install the hinges once for the function of the groom to be delegated forever to the cat-flap. That's the genius of a technical detour. A little time, a little steel, some screws, some sawing, and a function which made Prunelle a perpetual slave became the plan of action of a being which no longer resembles a human.

But, as for all innovations, there are conflicting interpretations. Prunelle thinks that there has been a destruction and not a new production: "Bravo! All the doors on this floor ruined!" To which the cunning Gaston retorts that at least Prunelle's rights to health have been respected: "but you have to admit that there are no drafts!" The cat-flap is a compromise: the delighted cat has ceased mewing; Prunelle, furious at first, will soon take satisfaction in the fact that he is no longer catching colds. The

Captions for the cartoon (numbering follows order of drawings):

1. *Cat:* Miaow! Miaow!
Prunelle: Coming, coming, phew!

2. *Prunelle:* . . . this happens more than twenty times a day. . . . I've become a doorman for cats! . . .

3. *Cat:* Miaow! Miaow!

4. *Gaston:* Come on!! Don't you know that a cat can't bear shut doors?! . . . and that it needs to feel free?! . . .
Prunelle: Is that right?! Oooh, the poor dear!!

5. *Prunelle:* Well, what I can't stand is drafts, and I'm telling you that these doors will be closed, ROGNTUDJ. . . . SCHLAM

6. *Gaston:* Each time Prunelle pesters me, I've found a trick to stop him being the strongest. . . .

7. *Prunelle:* Ah! Bravo! All the doors on this floor, ruined!!!
Gaston: But you have to admit that there are no drafts!

8. *Seagull:* IAAHHR!
Gaston: Oh no! It's jealous. . . .

9. *Prunelle:* RAAAH!
Seagull: HiHiHi HIAAHR!
Gaston: "Come on, don't be unfair: the door's closed; right or wrong?"

engineer's trick has managed to satisfy the mewing cat and the boss with the tender throat at the same time. (The "pest" in the word "pester.") Who paid the price of this negotiation? Doors. They have been ruined, redesigned, redefined. Gaston, despite his legendary laziness, has done a lot of work. And not forgetting the newspaper, *Spirou*, which finances this happy gang. A short detour, a small bill, and the crisis is resolved by technical bricolage which puts an end to the confrontation thanks to a compromise in which more non-humans are engaged. The dispute between cats and bosses is first displaced and then quieted by the adjunction of saws, screws, and hinges.

But the seagull has been forgotten! What's a seagull doing in a journalist's office? The origin of this Belgian peculiarity matters little to us here. Whatever the reason, the seagull, too, is complaining, and its cries are more piercing than the cat's. Its fury was not anticipated, and it threatens the fragile compromise that is holding together Prunelle, the cat, drafts, and cat-flaps. "laahhr," says the seagull. Gaston, a great animal psychologist, interprets the squawks as jealousy. Cats like to be free, and so do seagulls, especially when cats are. What should be done with this new and unforeseen actor crying out his fury and confusion? Eliminate it? Impossible: Gaston likes his seagull too much. Ask Prunelle to become a seagull-doorman, after he has refused to be one for the cat? Impossible. He'd fly off the handle. Offer the seagull the use of the cat-flap? The cat-flap is too tiny, and the seagull too proud to stoop to that level. Gaston has to take up his tools again, and go back to the doors to redefine them a little. "Try and try again" is the maxim of the inventor who has to bring the full weight of bosses, cats, and birds to bear on his inventions. He remakes them, redefines them. He adds a gap. Whoever invented the cat-flap could have invented the "seagull-gap," "a practical little opening at the top of a door that lets seagulls come and go," as the dictionary will soon say.

"Raah!" is all that Prunelle can manage to say. He was moaning figuratively before, and now he's really moaning, reduced to the mode of expression of cats and seagulls. Gaston, who understands animal language, takes Prunelle's groanings for an argument that he immediately counters with much good-heartedness: "Come on, you can't be serious: the door's closed; right or wrong?" Closed to drafts and open to cats and seagulls. Who could be unfair enough to pretend to the contrary? Who could be ass enough not to recognize a door—admittedly a renegotiated one—in the innovation offered by Gaston? When his apoplectic crisis is over, Prunelle will be forced to realize that the innovation pacifies all the crises and that the rights of cats, of seagulls, of flu-y bosses, and of errand-boy friends of the animals are all respected provided the door takes on certain modifications. The door bends itself, complicates itself, to take on the conflicts between people and animals. The cat-flap appeases the cat; the seagull-gap satisfies the seagull; the remainder of the door restrains drafts and should pacify Prunelle—so long as he's not really an insincere bastard, who, indifferent to technical invention, forces Gaston and his menagerie back to the door, to power, and to moanings.

No one has ever seen a technique, and no one has ever seen a human. We see only assemblies, crises, disputes, inventions, compromises, substitutions, translations, and orderings that get more and more complicated and engage more and more elements. Why not replace the impossible opposition between humans and techniques by association (AND) and substitution (OR)? Endow each being with a program of action and consider everything that interrupts the program as so many anti-programs. Draw up a map of alliances and changes in alliances. Maybe then we could understand not only Lagaffe but also Vulcan, Prometheus, Archimedes, and Daedalus.

It doesn't matter where you begin—that's what's interesting about this viewpoint—precisely because the assemblies mix things and people. Start, for example, with the cat. In version 3, everyone is against him, and Prunelle's fury serves him ill. But once the astute Gaston and his hinged cat-flap are counted among the cat's alliances the cat's plan of action can be fully realized. The cat doesn't even notice the difference between going through an open door, and going through a cat-flap. The translation becomes, for the cat, an equivalence: cat flap = open door = the freedom of the wild. As for Prunelle's anger (or that of the seagull), it no longer affects the cat. The irreversible cat-flap made of wood and hinges is immune to the mood swings of the cat-doorman. Completely indifferent, Gaston's cat goes all over the place as if nothing were the matter.[5]

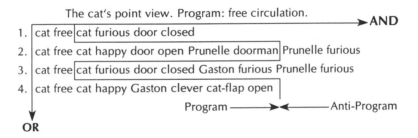

The cat's point view. Program: free circulation. ———————► AND

1. | cat free | cat furious door closed
2. | cat free cat happy door open Prunelle doorman | Prunelle furious
3. | cat free | cat furious door closed Gaston furious Prunelle furious
4. | cat free cat happy Gaston clever cat-flap open |

Program ————►◄———— Anti-Program

OR

The story is more complicated from Gaston's point of view because he has to reconcile several interested parties. The cat looks after only itself, and Prunelle cares only about his health and his newspaper. But Gaston has decided to keep everything around him; his animals, his work, and his bosses. Not wanting to renounce anything, he has to devise compromises between beings, things, and people. Not only must he redefine the door so that it incorporates first a cat-flap and then a seagull-gap, but he must also renegotiate Prunelle by offering him qualities that he seems not to possess. This is the big lesson of the philosophy of techniques: things are not stable, but people are much less stable still. Prunelle the journalist becomes a doorman. Prunelle is not a unity but a multiplicity. He is at one and the same time docile and exasperated, and it's on this multiplicity that Gaston is playing. From the bossy, grumpy Prunelle, Gaston imagines a Prunelle who will acknowledge that "there are no drafts," as Gaston says somewhat facetiously. Our Daedalus goes even further. He forces the apoplectic Prunelle to divide into one absolutely furious persona, and another pacified, sincere persona that recognizes the door that is open to all the animals as a good old closed door. Each redefinition of the door redesigns Prunelle's psychology and carries in its wake the acquiescence of the animals. There are as many Prunelles as there are doors and Gastons. There are as many doors as there are Gastons, Prunelles, and cats.

Gaston's point of view. Program: make everyone happy without having to choose between them.

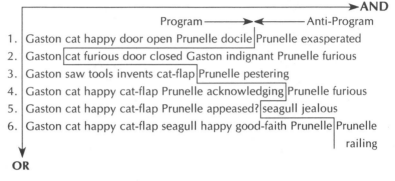

1. Gaston cat happy door open Prunelle docile | Prunelle exasperated
2. Gaston | cat furious door closed Gaston indignant Prunelle furious
3. Gaston saw tools invents cat-flap | Prunelle pestering
4. Gaston cat happy cat-flap Prunelle acknowledging | Prunelle furious
5. Gaston cat happy cat-flap Prunelle appeased? | seagull jealous
6. Gaston cat happy cat-flap seagull happy good-faith Prunelle | Prunelle railing

OR

One cannot do the philosophy of techniques without extending existentialism to the practical realm of inert things. Imagine a slightly more resistant Prunelle: solid as a rock, he would remain a pest; the ruined doors would have to be mended; he would refuse to accept that there were no drafts; out of his bad faith he would demand the departure of the farmyard. Imagine slightly more resistant doors: this time it would be Gaston that wouldn't be able to renegotiate them. Imagine more fragile animals: they would die at the first closed door they came across. If there were only essences, there would be no techniques.[6] Gaston insinuates himself into all the small existential fissures and tries multiple combinations until he finds one—almost at the point of apoplexy—which pacifies everyone in the little circle he has assembled around himself. The saw, on the other hand, and the toolbox and hinges are well entrenched essences that can serve as foundations. Likewise, the psychology of cats—"they like to feel free"—and of seagulls—"jealous"—which are not renegotiable. Essence is not on the side of things, and existence on the side of humans. The partition is between those who were an existence and provisionally become an essence—the resourcefulness of Gaston, the psychology of cats, the saw—and those who were an essence and provisionally became an existence—the psychology of Prunelle, the idea of a door.

Abandoning the false clarity of people facing objects needn't lead to chaos. On the contrary, it enables us to put to the test that which is possible and that which is not: the cat won't change its psychology, and Gaston won't abandon his cat; Prunelle will always be at risk of catching a cold and will always wish doors to be closed. In the place of distinct logics belonging to things of wood, flesh, or spirit, one can substitute so many socio-logics, more muddled perhaps, but less constraining: if the cat is made happy, then the seagull must be as well; if cat-flaps are installed in one door, then all the doors on the floor must be ruined; if Prunelle is happy then all the animals are crying out in frustration. Innovation is the test that permits the solidity of all these links to be tested. It's the trials of innovation, and they alone, that allow us to learn if the idea of a door is flexible, and if Prunelle is multiple.

Far from blurring distinctions, this philosophy actually permits one to disentangle the socio-logics. What is a technical innovation? Modifications in a chain of associations—numbered above from 1 to 6. Where do these modifications come from?

First, from the addition of new beings: the saw was no more expected than the cat-flap or the jealous seagull were anticipated.

Second, the passage of an actor from a program to an anti-program, or vice versa: the open door conspires with the cat but also with the drafts and so against Prunelle, for whom a breath of air is sufficient for him to catch a cold. An ally in a program of action becomes an adherent of the anti-program; something that conspired against the program becomes favorable to it.

Third is the change of state of an actor that finds itself endowed with new properties: the furious cat becomes happy, the jealous gull becomes happy, docile Prunelle becomes a pesterer, then furious, then unfair, then sincere, the classic door becomes Belgian, and Gaston finds himself ingenious instead of indignant.

Fourth, the modifications come from a substitution between beings: Prunelle cat-doorman is replaced by a cat-flap, a new assembly that prolongs the same function but in a different material.

Fifth, from a packaging, a routinization of the actors who have become faithful to each other: for the cat, all Gaston's work and the doors have disappeared, and the cat goes where it wants without noticing and purrs contentedly; for Prunelle soon (at least we hope so) work will reabsorb him as if nothing had happened with the new doors (open to cats, open to gulls, closed to colds). Fragile existences become stable essences once again, black boxes.

If we manage to follow these five movements and vary the point of view of the actor such that the same story mixes the cat, the door, the gull, the saw, Prunelle, and Gaston, then everything will be said. If the description is complete, the explication will soon follow: there exists one and only one door that can hold together the whims of Gaston, of Prunelle, and of their domestic animals. It may not be logically exact, but it's socio-logically rigorous. If we had focused simply on the evolution of this one door, as a historian of technology might have isolated it, we would have understood nothing about innovation, and nothing about conflicts in Belgian offices either.

Technical Evolution

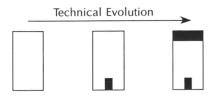

If the object is given to us with the beings that it holds in place and which hold it in place, we can understand the world in which we live. The door evolves by transpositions and substitutions, but Prunelle evolves too, and Gaston, and their animals. They don't evolve "in parallel" as is sometimes said, or by reciprocal influence, or by retroaction, or by dialectics. The door is like a word in a sentence, connected to other words. There is a single grammar for things and for people, and a single semantics.

Point of view of the door. Program: resolve the open/shut contradiction.

➤ **AND**

	Program ————➤◄———— Anti-Program
1.	door functioning shut │cat furious Prunelle hostile
2.	door functioning open cat happy│Prunelle furious
3.	door functioning shut Prunelle happy│cat furious Gaston furious
4.	door w/ cat-flap cat happy Prunelle acknowledging│Prunelle furious
5.	door w/ cat-flap cat happy Prunelle appeased?│seagull furious
6.	door w/ cat-flap seagull-gap cat happy seagull happy Gaston happy Prunelle made happy

OR

Contrary to the imaginings of the moralists, one cannot cut out the words of this long sentence without cutting out those that make up humanity. One can add actors and substitute certain of them by including others in a routine, but it is impossible to diminish the number: the door gets more complicated, Prunelle's psychology gets more complicated, the number of actors accumulates. Wanting to simplify these groupings by extracting the human actors, simplifying human essence, and placing humans face to face with things equally reduced and isolated is a barbaric form of torture which, I hope, will no longer be paraded under the beautiful name of humanism.

Notes

1. This expression derives from the title of a nineteenth-century play by A. de Musset, *Il faut qu'une porte soit ouverte ou fermée: proverbe en 1 acte* (Paris: Librairie Théatrale, 1848), whose moral is that one must decide one way or another to avoid the incapacitating effects of indecision. The English proverb "fish or cut bait" carries the same or similar significance. It is used in this piece by Latour to suggest that perhaps one *can* have one's cake *and* eat it.—Translator's note.

2. See Bruno Latour, "Where Are the Missing Masses? Sociology of a Door," in W. Bijker and J. Law (eds.), *Shaping Technology—Building Society: Studies in Sociotechnical Change* (Cambridge, MA: MIT Press, 1992).

3. F. Frontisi-Ducroux, *Dédale, Mythologie de l'artisan en Grèce Ancienne* (Paris: Maspero, La Découverte, 1975).

4. Plutarch.

5. B. Latour, P. Mauguin, and G. Teil, "Une méthode nouvelle de suivi des innovations. Le Chromatographe," in D. Vinck (ed.), *La Gestion de la recherche: Nouveaux problèmes, nouveaux outils* (Brussels: De Broek, 1991).

6. "Objets techniques": see G. Simondon, *Du mode d'existence des objets techniques* (Paris: Aubier, 1989).

Contributors

Albert Borgmann is Professor of Philosophy at the University of Montana. His publications include *Technology and the Character of Contemporary Life* and *Crossing the Postmodern Divide*.

Hubert L. Dreyfus, Professor of Philosophy at the University of California, Berkeley, is author of *Being-in-the-World: A Commentary on Heidegger's* Being and Time, *Division I, What Computers* Still *Can't Do*, and coauthor (with Paul Rabinow) of *Michel Foucault: Beyond Structuralism and Hermeneutics*.

Paul Dumouchel is Professor of Philosophy at the University of Quebec. He has published numerous articles on philosophy of science and technology, political philosophy, and philosophy of mind, and is editor of *Violence and Truth*, author of *Le corps social: essai sur les émotions*, and coauthor (with J. P. Dupuy) of *L'enfer des choses*.

Yaron Ezrahi is Professor of Political Theory and Democratic Institutions at the Hebrew University of Jerusalem and author of *The Descent of Icarus: Science and the Transformation of Contemporary Democracy*. His work covers a wide spectrum of subjects in the field of science, culture, and politics in the modern democratic state.

Andrew Feenberg is Professor of Philosophy at San Diego State University. He has written widely on philosophy of technology and on the application of computers to human communication. His books include *Lukács, Marx, and the Sources of Critical Theory* and *Critical Theory of Technology*.

Alastair Hannay is Professor of Philosophy at the University of Oslo. His books include *Mental Images: A Defence, Kierkegaard*, and *Human Consciousness*. He is editor of the journal *Inquiry*.

Donna Haraway is Professor of History of Consciousness at the University of California at Santa Cruz, where she teaches feminist theory, cultural and historical studies of science and technology, and women's studies. She is author of *Crystals, Fabrics, and Fields: Metaphors of Organicism in Twentieth-Century Development Biology, Primate Visions: Gender, Race, and Nature in the World of Modern Science*, and *Simians, Cyborgs, and Women: The Reinvention of Nature*. She is currently completing a book entitled *Worldly Diffractions: Feminism and Technoscience*.

Marcel Hénaff is Professor of Literature at the University of California, San Diego. An English translation of his *L'Invention du Corps Libertin* is scheduled for publication by University of Minnesota Press.

Don Ihde is Leading Professor of Philosophy at SUNY Stony Brook. His publications include *Postphenomenology, Philosophy of Technology, Technology and the Lifeworld*, and *Instrumental Realism*.

Bruno Latour, a trained philosopher and anthropologist, is a professor at the Centre de Sociologie de l'Innovation in the Ecole Nationale Supérieure des Mines in Paris. He is author of *Laboratory Life: The Construction of Scientific Facts, Science in Action: The Pasteurization of France, La clef de Berlin, Aramis or the Love of Technology*, and *We Have Never Been Modern.*

Helen E. Longino, Professor of Philosophy at Rice University, is author of *Science as Social Knowledge* and numerous articles in philosophy of science and feminist philosophy. Her research interests include philosophical issues in biology, the social dimensions of scientific knowledge, and feminist theory.

Robert B. Pippin is Professor of Social Thought and Philosophy at the University of Chicago. He is author of *Kant's Theory of Form, Hegel's Idealism: The Satisfactions of Self-Consciousness*, and *Modernity as a Philosophical Problem: On the Dissatisfactions of European High Culture.*

Tom Rockmore is Professor of Philosophy at Duquesne University. His recent publications include *Habermas and Historical Materialism, On Heidegger's Nazism and Philosophy, Lukács and the Marxist View of Reason, Before and After Hegel, Hegel et la tradition allemande*, and *Heidegger and French Philosophy: Humanism, Anti-Humanism, and Being.*

Pieter Iijmes is in the Department of Philosophy and Social Sciences at the University of Twente, Netherlands. He has written on the role of science and ethics in the political thought of Max Weber, and is currently doing research in the philosophy of technological culture.

Steven Vogel is Associate Professor of Philosophy at Dennison University. He is the author of *Nature and Critical Theory.*

Langdon Winner is Professor of Science and Technology Studies at Rensselaer Polytechnic Institute and author of numerous books and articles, including *Autonomous Technology, The Whale and the Reactor*, and "Do Artifacts Have Politics?"

Terry Winograd is Professor of Computer Science at Stanford University. He has written extensively on artificial intelligence, and is coauthor (with Fernando Flores) of *Understanding Computers and Cognition: A New Foundation for Design*, and coeditor (with Paul Adler) of *Usability: Turning Technologies into Tools.*

Index